The 1976 Tangshan Earthquake:

Memoir and Record of Disaster Preparedness and Mitigation

Sun Zhuozi
Chief Editor

Casey Huang
Editor and Translator

The 1976 Tangshan Earthquake: Memoir and Record of Disaster Preparedness and Mitigation
Chief Editor: 孙卓梓 (Sun Zhuozi)
Editor and Translator: 黄可欣 (Casey Huang)

This book is a revised international edition of a work originally published in Chinese.

First Australian Edition 2026

ISBN: 978-0-646-73876-5
Publisher: Inkcarve press

Cover design by 施沁怡 (Qinyi Shi)

Preface

Dear readers,

We are survivors of the 1976 Tangshan Earthquake, one of the most devastating earthquakes of the twentieth century. Even after all these years, the terror of that night and the hardship that followed remain vivid in our memories. We have long felt a responsibility to share what we witnessed and endured.

This book tells not only our personal stories, but also what we have learned from that catastrophe. We describe the unusual natural phenomena that can sometimes precede earthquakes, how to protect oneself during an emergency, and how to cope with difficult conditions in the aftermath. Our hope is to pass this knowledge on to younger generations, so that they may be better prepared should disaster strike. With their lives still stretching far ahead, it is important that they understand both the risks and the responsibilities of preparedness.

This volume has been prepared with the guidance and support of the Ministry of Education, the Ministry of Science and Technology, the Tangshan Emergency Management Bureau, the Tangshan Earthquake Prevention and Mitigation Bureau, the Tangshan Earthquake Memorial Hall, and Hebei Provincial Geological Survey Team No. 2.

Drawing on representative stories from Forty Years of New Tangshan: The Rebirth of Tangshan People After the Earthquake, this book combines scientific knowledge with the lived experiences of survivors. It is intended to help students and their families understand practical strategies for disaster preparedness, so that when earthquakes or other emergencies occur, they can respond calmly and wisely.

Finally, we extend our sincere gratitude to all those who supported and contributed to the publication of this book.

Chief Editor
28 July, 2023

Dedication

In Commemoration of the 50th Anniversary of the Tangshan Earthquake

Dedicated to those who lost their lives in the earthquake, those who survived it, and those who continue to dedicate themselves to disaster preparedness and mitigation.

Table of Contents

Tangshan Earthquake
Date: 28 July, 1976
Location: Tangshan

The Tangshan Earthquake refers to a powerful earthquake of magnitude 7.8 that struck Tangshan and Fengnan district on 28 July 1976. The seismic intensity reached level XI. More than 242,000 people were killed, and about 164,000 were seriously injured. The earthquake also affected Tianjin and Beijing, with a vast impact area covering 14 provinces, municipalities, and autonomous regions. The disaster completely devastated Tangshan, a major industrial city with a population of over one million, and caused economic losses of approximately 10 billion RMB. It is regarded as one of the ten major natural disasters of the 20th century.

28 July 2026 marks the 50th anniversary of the 1976 Tangshan Earthquake.

On the occasion of this solemn commemoration, and with the strong support of the Tangshan Municipal Party Committee and Municipal Government, a collection of memoirs was compiled under the editorship of Sun Zhuozi. A total of 270 earthquake survivors contributed their personal recollections, offering first-hand accounts of the devastating Tangshan earthquake of 1976.

These memoirs faithfully record what the survivors experienced before the earthquake, how they struggled to survive and escape amid the destruction, and how they endured the most difficult period of life in the aftermath. Many of these contributors have since passed away. Yet through their words, they have left behind the most precious memories and testimony for the world, living evidence of resilience, perseverance, and endurance in the face of catastrophe.

28 July 1976, Tangshan 7.8-magnitude Earthquake

Chapter One

Memories Before the Earthquake

Two or three months before the earthquake, rumours began circulating among the residents of Tangshan that a major quake might be coming. Some families placed empty beer bottles upside down on the floor each night, believing that if an earthquake struck, the sound of the bottles falling would serve as an alarm.

My family lived in a single-storey house[1] with a paper-covered roof. About two months before the earthquake, strange noises began every night as soon as the lights were turned off. The roof would rustle loudly, as if rats were scurrying about, or as if swarms of insects were crawling back and forth. The moment the lights were switched on, the sounds would stop completely. The noise was so disturbing that it made it difficult to sleep.

On the night before the earthquake, the weather was unusually hot, and I felt extremely restless and irritable. I had been granted sick leave that day and did not need to go to work, but I felt so uneasy that I got on my bicycle and went to work the night shift anyway, despite being ill.

— Mr Sun

National Key Cultural Heritage Protection Site
Tangshan Earthquake Ruins
Proclaimed by the State Council, May 2006
Established by the People's Government of Hebei Province

[1] *In China, single-storey houses built mainly with timber and mud-brick were common before the 1980s.*

On the afternoon of 27 July 1976, I accompanied provincial and regional leaders as we drove south from Luan County along the Luan Rive embankment to inspect flood-control work. As our vehicle passed Macheng in Luan County, the Luan River, which normally flows quietly to the south, seemed to have suddenly grown turbulent. The water level appeared to rise sharply in a short time, and as we continued south, the river kept swelling.

When we returned to the Prefectural Office building, we learned that the upper reaches of the Luan River had experienced several consecutive days of clear weather and that there were no abnormal hydrological reports. A troubling question then settled deeply in our minds: if no rain had fallen in the Chengde region, where had this surging river water come from?

— Mr Meng

That night was unusually stifling. As my father later recalled, he felt so hot that he could not fall asleep. He lay there fanning himself with a palm-leaf fan for most of the night. Eventually, he got up and went to the east room.

At that time, our house consisted of three tiled rooms, with a central hall in the middle and living rooms on either side. Our family slept in the west room, while the east room had never been renovated. It was used to store odds and ends and was not lived in, so it was relatively cooler. After going into the east room, my father found a flat spot and lay down. He later remembered that not long after he lay down, the earthquake struck.

— Mr Zhao

On 27 July 1976, the teacher training program at Kailuan No. 2 Middle School came to an end, and that evening marked the first night shift of the school's summer security duty. I was responsible for campus security. That night, together with the on-duty security staff, I inspected key areas of the school for potential safety risks.

When we reached the crossroads on campus and stopped beneath a large locust tree with thick, spreading branches, we suddenly heard the sound of wings beating, fluttering above us, followed by a series of low hoots. The cry of an owl in the middle of the night was deeply unsettling. Even after nightfall, the heat remained intense, like being inside a steam bath. When I rubbed my arm, small rolls of dirt came off. The air was so muggy that it was hard to breathe. We returned to the duty room, where Vice Principal Gao Tianzhang said to me, "Xiao Guo, go back and get some sleep. Aren't you heading back to your hometown at daybreak?" I looked up at the wall clock. It was already 2:45 a.m.

— Mr Guo

In April of the year the earthquake struck, I was transferred to the Tangshan Air Force Station, where I lived in the main administrative residential compound. On 27 July, the weather was stifling. Not a single leaf stirred. The air felt thick and motionless, as if it had solidified. That night, I tossed and turned, unable to fall asleep. I do not know how much time passed before I drifted into a hazy doze. Suddenly, my bed began to shake violently. I opened my eyes and saw flashes of light flickering outside and I heard rolling thunder. My first thought was that enemy missiles were incoming.

At the time, international tensions were high, and we lived in a state of constant readiness, the string of war drawn tight. I immediately slid under the bed for cover. Then, in a split second of lightning, I dashed out of the room. No sooner had I reached the doorway than a deafening crash split the air. The building I had been living in collapsed behind me. It was a narrow escape. Had I stepped out even a few seconds later, I would have been gone from this world.

— Mr Kou

Summary of this relic

On 13 July 1976, we arrived at a small village to join the Down to the Countryside Movement[2]. The village was very small, with only a little over thirty households. When more than a dozen sent-down youth[3] arrived all at once, the village seemed too crowded to accommodate us. The production brigade arranged for us to be spread out and lodged with local families.

Four classmates and I were sent to a household on the eastern side of the village. There were only a mother and her son in the family. The elderly woman looked kind and gentle, often watching us quietly. Her younger son was solidly built, about twenty-six or twenty-seven years old, and served as a small team leader in the production team. For some reason, he always lowered his head and spoke very little whenever he saw us. The young people in the team all called him Er Ge (Second Brother), and following local custom, we did the same. By coincidence, we were assigned to his team.

On 27 July, the weather was exceptionally hot. That evening, after returning from work, we were so exhausted that we lay down on the kang[4], saying nothing, and were too tired even to eat. Liu Shiyou nudged me and said, "it is your turn to carry water again." Very unwillingly, I got up, picked up the

---

[2] *A movement during China's Cultural Revolution in which many urban youths were relocated to rural areas to work and learn from peasants.*

[3] *Sent-down youth were urban young people sent to rural China to live and work under the Down to the Countryside Movement.*

[4] *A kang is a traditional heated brick platform bed commonly found in northern China, warmed by flues carrying heat from a stove beneath it.*

buckets, and went to the well to fetch water.

To make washing more convenient, we had agreed that one person would fetch water each day. I was first, followed by Cui Yufeng and Chen Xiangdong. Four young men in their early twenties were crowded into a small room, and in such heat, we used a lot of water.

When I reached the well, I found something strange. The well, which was usually very deep, seemed unusually full that day. The water level was so high it looked as if it might overflow. I did not even need to use the carrying pole. By simply turning the bucket upside down, it filled completely. When I carried the water back, I told the others about this odd situation, but no one paid much attention. Liu Shiyou said that perhaps there had been continuous rainfall over the past few days, causing water to flow back into the well. I replied that if that were the case, the water should have been very muddy, but in fact, the water I drew was very clear. Liu Shiyou thought for a moment and said we should go ask Second Brother. Second Brother was outside the courtyard, standing under a willow tree, trying to drive a chicken down from its branches. The landlady stood nearby, calling to the chicken repeatedly. The chicken perched in the tree and refused to come down no matter what. Already in a bad mood, Second Brother turned and saw the two of us, then asked what we wanted. I explained what had happened. Frowning, Second Brother said that people from other teams had also mentioned the water level in the wells rising over the past few days. The elders had all said they had never seen anything like it before. Perhaps, he added, it had something to do with the heavy rain.

— Mr Shi Youshan

Chapter Two

Escape During the Earthquake

In 1976, a magnitude 7.8 earthquake struck Tangshan, Hebei Province. At the time, I was working at a textile factory and was on the night shift.

It was just after three in the morning. As usual, we were preparing to make breakfast in the workers' canteen. The workers had brought dried noodles and eggs, placing them in their lunch boxes. I carried them to the canteen together with the shift leader.

At around 3:40 a.m., the food had just been cooked, but before we had any chance to eat, a terrifying sound suddenly came from the east, moving rapidly toward the west. It was a deafening, crackling roar, like rolling thunder, growing closer by the second. I instantly realised that it was an earthquake. I had barely taken two or three steps forward when the ground, surging like a massive wave, knocked me flat.

I saw a table tennis table nearby and crawled desperately beneath it. At that moment, wooden boards, asbestos tiles, and bricks came crashing down from above. The shaking lasted for about five minutes. When the ground finally grew still, I saw light coming through the ruins. I quickly pushed aside the wooden frames, bricks, and asbestos tiles and climbed out. From the beginning to the end, no more than ten minutes had passed.

What I saw before me was only debris. Not a single building remained standing. In the distance, two columns of black smoke were rising into the air. A chilling thought crossed my mind: Could I be the only one left alive in Tangshan? Within minutes, cries for help began to break out from all directions. The shift leader was still buried beneath the rubble. I rushed back, cleared away the debris, and pulled him out. Together, we moved along the factory offices and dormitories, digging through the ruins and rescuing people. We went on for hours, pulling out everyone who could still call for help, until it was close to ten that morning. By then, I was completely exhausted. I collapsed in a spinach field beside the factory, tore up a handful of spinach, and ate it raw. Then I went back to rescue more people.

Looking back now, I am convinced of one thing: I was the first person in Tangshan to stand up from the rubble of the great earthquake.

— Mr Sun

In the early hours of 28 July, a time that should have belonged to the deepest sleep, I was still wide awake. I lay there turning the pages of this booklet, utterly absorbed. My wife stirred, glanced at the clock by her pillow, and said, "It's already past three. Why aren't you asleep yet?" I didn't answer. A passage I was reading, quoted from *The Qianlong Annals of Sanhe County*, had completely seized my attention: "On the twenty-eighth day of the seventh month in the eighteenth year of the Kangxi reign, an earthquake struck the capital. Rising from the northwest, sand and dust filled the sky, plunging it into darkness. People were thrown about as if on rolling waves, unable to remain upright. Thunderous sounds followed, sending birds and beasts into panic. The ground shook repeatedly through the night, splitting open the earth and tearing a great fissure beneath Desheng Gate, where water surged up like a spring. The dead and wounded among both officials and civilians were beyond counting; bodies lay piled upon one another, beyond recognition."

I was reading intently when my wife urged me again, "Go to sleep. Don't you have to go out of town tomorrow?" As she spoke, she turned over and held our younger son over the potty. At that very instant, a flash of blue light lit up the entire night sky outside the window. A thunderous roar followed.

The ground shuddered violently, and the dust poured down from the ceiling. Frightened out of her wits, my wife loosened her grip. With a dull thud, our son dropped into the potty. I sprang to my feet and shouted, "Earthquake!" The shout was so loud that, had it not been swallowed by the roar of collapsing buildings, it would have jolted everyone in the building awake. My wife cried out in panic, "Go! Open the door!" I yanked our son out of the potty and shouted back, "Don't get off the kang. Run to the window!" I knew that during an earthquake, the sudden air pressure makes doors nearly impossible to open.

After that flash of blue light, cries of terror came from every direction. Then, after a single earth-shattering boom, the entire city fell into a deathlike silence. Moments later, heart-rending screams erupted from all sides, tearing through the dark. Clutching our son, my wife cried out, "I'm not even wearing my trousers!" I snatched the child under my right arm and cursed, "What the hell do you need dignity for now? Even naked, you have to run!" Later, facts would prove this decision to be remarkably wise. In the days that followed, some women who were pulled from the rubble were found still clutching their trousers in death. Sadly, some lost their chance to escape simply because they were trying to get dressed.

Holding our son and dragging my wife along, I had just reached the south-facing window when a tremendous crash sounded behind us. The wall by the door collapsed inward toward the kang. We had narrowly escaped disaster. If I had followed my wife's idea and climbed down to open the door, we would have been smashed to pulp. Amid the roar of collapsing houses and endless screams, I kicked hard at the glass window. With a loud crash it shattered, but the jagged frame left no space to squeeze through. I was trapped, unable to go forward or retreat, sweating in desperation. Then the ground shook violently again. Then the ground convulsed violently again. A crack more than a foot wide split open between the west wall and the window frame. Our family of three squeezed through the gap and stepped onto the outer windowsill. Below the sill was a three-foot-wide passageway, beyond which stood a small shed I had built to store firewood, coal, and odds and ends. Its brick walls and solid roof made it exceptionally sturdy. Just as we were about to jump down from the sill, the shed suddenly collapsed with a hollow roar. Broken bricks and stones piled up level with the windowsill, perfectly forming a path of escape.

I hurried onto the rubble of the collapsed shed. Just as I was about to step forward, the south wall behind us and the window came crashing inward with a heavy thud. I felt a quiet sense of relief. Had we been even one second slower crawling out of the room, we would have been crushed beneath the falling wall. Had we jumped from the windowsill one second earlier, we would have been smashed by the shed outside. I still cannot tell whether this was fate or mere chance. When our family of three reached Xinhua Avenue in front of the building, the wide road lay completely empty. Beneath a concrete utility pole by the roadside, only two figures were moving. It was Li Lianzhong from the newspaper office and his wife. Of the entire newspaper building, they were the first to escape. We were the second.

In truth, from the moment the blue light flashed to the instant I was out of danger, no more than three to five seconds had passed.

— Mr Meng

I still remember the early hours of July 28. People were fast asleep when a sudden gale seemed to rise from nowhere. A strange blue light flickered. It felt as if the sky and earth were splitting apart and the earth turning over. I was jolted awake and sat up abruptly. Instinctively, I scooped up my daughter, who had just turned one. I turned to wake my husband. I caught only a glimpse of him before he vanished from sight. In the next instant, everything fell into darkness, and our whole family was buried in the rubble.

In the darkness, I could still feel the earth trembling. Terror overwhelmed me. Holding my daughter tightly, I curled myself around her, frozen and helpless. When the shaking finally stopped, a small opening appeared above and to our right. A thin beam of light shone through. I was suddenly filled with hope. I quickly clawed the opening wider and crawled out. By sheer luck, my daughter and I made it out of the tangled wreckage with only minor scratches.

Standing on the rubble, clutching my child, I trembled uncontrollably and shouted for help. "Help us! Her dad is still buried in there!" Hearing my cries, my mother-in-law rushed over from the neighbouring house. She took the child from my arms and urged me, "Go dig for his father!" I stared at the ruins, my legs shaking, not knowing where to begin. At that critical moment, my two sisters-in-law ran over. Together they lifted a massive slab of the collapsed roof. I steadied it from the side as they climbed inside, calling out my husband's name again and again. After several minutes, a faint response came from within. One of my sisters-in-law cried out in relief, "He's here!" My husband was completely buried beneath bricks and earth. His exact position was impossible to see. My two sisters-in-law dug frantically with their bare hands. When their hands were already torn and bleeding, they finally uncovered him. When he was pulled out, his face was dark and badly swollen, his body burning with heat. He could not stand on his own. With everyone's help, he was dragged from the rubble, his life narrowly saved.

— Mrs Wang

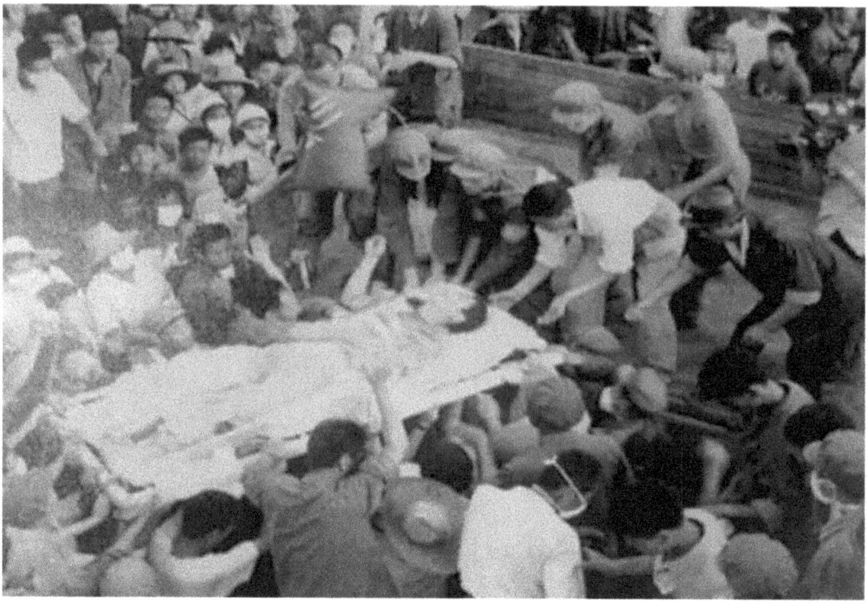

I was half-asleep when I suddenly heard a strange sound, something like a distant *rumble... rumble*. It seemed to come from far away, yet at the same time from deep beneath the ground, low and frightening. *Rumble... rumble...*

I jolted upright instinctively. The kang beneath me bucked violently. My husband and I jumped to the floor at the same moment. He shouted, "Earthquake!" and reached for the light. The bulb flared white like a flash of lightning across the dark room, then went out with a sharp snap. In that split second, I saw the walls quivering like sheets of paper. I screamed. At the same time, I grabbed the legs of our two children and dragged them off the bed. Before I could pull them under me, the house collapsed with a deafening crash.

I was frighteningly clear-headed. I reached out but could not find my kids. Thick smoke and choking dust filled the air, swirling like a dense fog. I could see nothing around me, and I could not see my husband. My heart sank. The sudden disaster left my mind completely blank. I thought: "Where am I? Where is our house?" Just then, another explosion thundered through the darkness. Red flames tore through the haze. A shop not far from our home was on fire, and the fire ignited the oil drums stored in the shop. By the firelight, I forced myself to calm down. My husband and children were nowhere in sight. As far as I could see, every house had collapsed.

Where are they? I began calling my husband's name. At first there was no answer. After calling several times, I finally heard his weak reply: "I'm by the door." I steadied myself and realised that I was standing on what remained of the roof. My husband and the children were buried beneath the rubble. The roof under my feet, about half a foot thick, had already split into four pieces. In the darkness, I caught sight of a man's silhouette to the east.

"Please help me!" I shouted.

"Don't come over," he called back. "I'm not wearing any clothes."

"I don't care," I shouted. "Just help me save them!"

No matter how much I pleaded, he would not come. In desperation, I picked up a rafter as thick as an arm from the rubble and thrust it beneath the slab pinning my husband down. I leveraged it with all my strength. The slab did not move at all. Instead, the timber snapped in two. Long after the earthquake, I realised that I could never have had such strength under normal circumstances. It was the extremity of the moment that forced it out of me. When the rafter broke, I pulled out another beam like a madwoman and wedged it beneath the roof slab again. I shouted once or twice more, hoping someone might help us, but there was only dead silence.

"Is anyone there?" my husband called.

"There is someone," I answered, "but he won't come no matter what I say."

"Then stop shouting," he said. "Save your strength. You'll need it to dig us out."

I stopped calling. Under his instructions, I pried the roof up little by little, propping it higher each time with bricks. When it was raised enough, I finally saw my husband and our son. My son's upper body was exposed, his lower body still buried, but he was breathing. I lunged forward and tried to pull him free, but it was useless. "Mum," he said, "don't worry about me. I can still breathe. Get Dad out first." At that moment, a violent aftershock struck. The ground rolled beneath me like a small boat caught in heavy seas. Barely able to keep my footing, I staggered to my husband and saw that a stone slab lay across his neck and shoulders, as large as a school desk. There was no time to think. I began clawing at the earth, fragments, and stalks of sorghum used in the roofing with my bare hands. At first, I heard my husband urging me urgently, "Hurry! Hurry!" After those two cries, his voice began to fade. Panicked, I reached blindly into a gap and grabbed hold of his hair. Summoning all my strength, I pulled hard and dragged his entire body free. He gasped for air. "I almost suffocated." Only then did I realise he was soaked through, sweat and dust mixed into mud all over him. Seeing that he was safe for the moment, I rushed back to our son. By then, my son had freed one arm with the other and had spotted his little sister beneath the rubble. He cried out, "Big Bubble!" I ran over and saw my daughter. A huge blister on the back of her neck rose and fell as she struggled to breathe. A deadly slab of stone pressed down on her head. I was almost despairing, but I tried the same method again. I thrust my hands under the slab and pulled with all my might. She came free.

— Mrs Shen

At around half past three in the morning, the baby began to cry. My wife got up to change the nappy and feed him. At that moment, she heard a few dull thuds, *thump, thump*. She nudged me and said, "Listen! There's noise outside. Do you think someone's knocking at the gate?" I tilted my head and listened. All I could hear were chickens crowing and dogs barking. The noise was constant and chaotic. I wondered why the chickens and dogs were making such a racket in the dead of night. My wife finished feeding the baby, turned off the light, and lay back down.

Just then, the house began to shake violently. The beams, jars, doors, window frames, cupboards, mirrors hanging on the wall, and the bottles stacked on top of the cabinets crashed into one another in wild confusion. My wife screamed, "What's happening?!" I realised instantly that it was an earthquake and shouted, "Earthquake!" She cried, "Grab the baby!"

I leapt off the kang and turned back to pick him up. My hand had barely reached out when the thick earthen wall on the west side of the house collapsed, burying the three of us beneath it. Immediately afterwards, the beams and the rest of the house came crashing down as well. My legs were pinned below the edge of the kang, my body bent at a right angle across it. My wife was pressed face down, her head toward the kang. None of us had managed to reach the baby.

After the first violent shock passed, there was a brief moment when everyone was still stunned and had not yet come to their senses. All around us, it fell eerily quiet. I mustered every bit of strength I had, braced myself on my elbows, and tried to heave the weight off my body. I tried again and again, to no effect, it would not budge. I thought, if we couldn't get out, the baby had no chance of surviving. Despair overwhelmed me. I began shouting, "My child! My child! Help! Help!" My wife, however, was calmer than I was. She said, "Stop shouting. You'll run out of strength, and they won't be able to hear you outside."

Aftershocks continued to strike, and the weight pressing down on us grew heavier. Our breathing became shallow. At that critical moment, we suddenly heard my brother-in-law calling out, "Sister, where are you?" We answered at once, "We're here!" But outside it was too noisy. All around us, people were shouting as they searched for loved ones and cried for help. They couldn't hear us. "Sister, where are you?" he called again. We shouted back repeatedly, "We're here! We're here!" At last, they heard us. Locating our position, they began clearing the debris by hand, layer by layer. After about ten minutes, the weight above us was removed, and our heads and hands were freed. It turned out that my father-in-law's house had not completely collapsed. They had managed to crawl out and came straight to rescue us.

"Where's the baby?" they asked.
"We don't know."
"Then go dig for the baby," they said. "We'll help the neighbours."

— Mr Xu

Second by second, time dragged on. By dawn, under a dim and overcast sky, rain began to fall. My father and I struggled out of the rubble and climbed onto the collapsed roof of our house. All around us were fallen buildings. In the distance, there were flames and drifting clouds of smoke. Aftershocks kept coming. Survivors were crawling out of the ruins one after another.

People were wearing nothing but their underwear. Their faces and bodies were caked in dust. Some were wounded, blood running down their faces and limbs. Cries for help kept rising from the rubble. It was a terrifying scene, one that can never be forgotten.

The rain continued. My father found a towel blanket in the debris. I wrapped it around my little sister and held her tightly as she cried in fright, calling out, "Mum! Mum!" My whole body trembled. I did not know what to do. My father stood nearby, his face filled with pain and bewilderment. Suddenly, he remembered Uncle Zhang and Aunt Zhang from the front courtyard. He began calling their names again and again. After some time, a faint cry for help came from the ruins of the collapsed house in their courtyard. Following the sound, my father cleared away the rubble bit by bit with his bare hands and pulled Aunt Zhang out from beneath the fallen roof. Her right leg was injured and bleeding. My father tore a bedsheet into strips and made a simple bandage for her leg. What about Uncle Zhang? Only then did we learn from Aunt Zhang that he had gone fishing that night. The heat had been oppressive, and he had been unable to sleep. He was not at home when the earthquake struck.

After settling Aunt Zhang and asking me to look after my little sister, my father joined the surviving neighbours along the street. Men and women alike came together spontaneously to rescue others. A cry for help became a command. Wherever there was a cry, people went to save whoever was trapped there. At that moment, everyone shared the same resolve. Even though their own family members might still be buried beneath the rubble, their fate unknown, they put all their effort into saving those whose voices could still be heard. Together, they searched for any tools they could find, such as shovels and wooden poles, to clear the debris. When there were no tools, they used their bare hands. In the rain, survivors were pulled out of the rubble one by one. Those who were young, strong, and not seriously injured joined the rescue efforts as soon as they were freed. Uncle Zhang, who had been fishing outside the town, rushed back and immediately joined the rescue as well.

By around three o'clock in the afternoon, all the survivors on our street who could still call out had been rescued. The rain finally stopped. Families carried the severely injured on makeshift stretchers made from doors to the main road to seek help. Some were taken to the Tangshan military airport to await further rescue.

— Zhiqiang

In 1976, I was teaching at Tangshan No. 1 Middle School. In the early hours of 28 July, after the ground lurched violently, I was fortunate to survive. I remember that when I first crawled out of the rubble, it was still pitch-dark. A fine drizzle was falling from the sky. Groans and cries for help rose again and again through the choking dust. Dazed and at a loss, I suddenly heard my name being called clearly. I looked up and saw that the voice was coming from the direction of the physical education office, only a short distance away. All around it, rows of classrooms and dormitories, along with the tall teaching building, had vanished in an instant.

An earthquake. A devastating earthquake. Without the slightest warning, it had descended mercilessly upon our city, destroying the beautiful campus beneath my feet. There was no time to think. Saving lives came first. I ran straight to the physical education office. Large slabs of the cinder roof had fractured into countless fragments and collapsed to the ground, completely burying several single-storey rooms. I was still young at the time, and perhaps in moments of extreme crisis people are capable of extraordinary strength. Before long, I managed to lever away one slab after another and pull out Mr Wang Shuoyun, who was also strong and well built.

Strangely enough, Mr Wang was completely uninjured. He brushed the dust from himself and saw that blood was running from my head and feet, while I had nothing on at all. He quickly tore a few pieces from a broken mosquito net, wiped me down, and tied the net around my waist. There was no time to waste. After a brief exchange, the two of us ran towards the next cries for help. By then, daylight had broken, and the rain continued to fall steadily. With two people working together, the rescue went much faster. I do not know how much time passed before we finally freed Mr Shen Hanxi, a physics teacher. He waved his hand weakly and told us that his wife and child beside him had already stopped breathing, urging us to go and save others at once. Fighting back tears, we left Mr Shen and rushed to another rescue site, where we dug out Mr Ma Ruishan, who had been buried for several hours.

In the end, we circled the campus once more. When we could no longer find any place where cries for help were coming from, we collapsed to the ground in exhaustion and sat facing each other. Only then did we realise that both of us were covered head to toe in blood, mud, and grime, like wild men. Those who had escaped death gathered instinctively on the open sports field. Looking into the frightened faces around me, I became acutely aware that the school's losses in this sudden catastrophe must have been enormous. Subsequent figures confirmed this. Because the school lay in the area of greatest destruction, 105 people died on the campus that day alone, far outnumbering those who survived. More than 800 school buildings collapsed almost entirely, and over 50,000 books, along with vast quantities of teaching equipment, were buried beneath the rubble.

— Mr Zhang

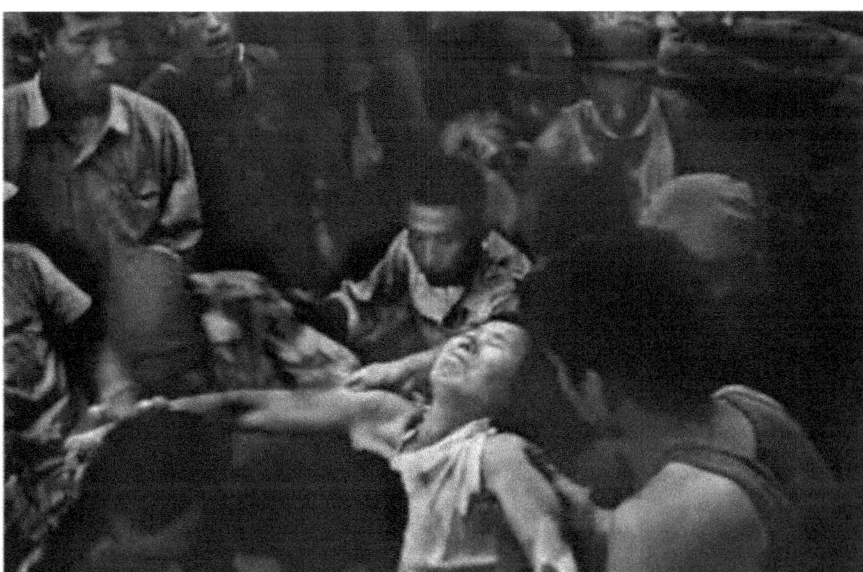

Chapter Three

Life and Survival After the Earthquake

After the Tangshan earthquake, the struggle for survival became the first and greatest challenge facing every Tangshan resident who made it out alive. There was no water, no electricity, no shelter, no transport, no grain, no vegetables, and nothing to cook with.

This is how I lived through the extraordinarily difficult week that followed the quake.

On the first day, I searched through the factory ruins, digging people out and helping with rescues. When hunger set in, I pulled spinach straight from the ground and ate it.

On the second day, word spread that there would be major flooding in Tangshan after the earthquake. I followed the crowds on foot to Fengrun. At that point, I barely felt hunger or thirst. That night, I slept outdoors in the courtyard of the Fengrun Hospital.

On the third day, I returned to Tangshan to search for my elder sister's family in a workers' residential block. I found three bricks, uncovered an iron pot, and dug out a bag of millet. My elder brother and I cooked a small pot of millet porridge. I have never known anything to taste so good.

On the fourth day, I dug out several plastic bags, cut them open, and made a sheet of plastic. By the roadside, I built a small shelter and found a few wooden planks to use as a bed. Lying down on them felt indescribably comfortable.

On the fifth day, a full company of the People's Liberation Army arrived to carry out rescue work. The workers' residential block was made of massive concrete slabs that could not be moved by hand. The soldiers smashed them apart with sledgehammers. Around this time, military units began delivering water and compressed biscuits. After about a week, aircraft were making daily airdrops of flatbread and steamed buns. In that moment of crisis, the People's Liberation Army played a crucial role in delivering emergency supplies to the people of Tangshan.

— Mr Sun

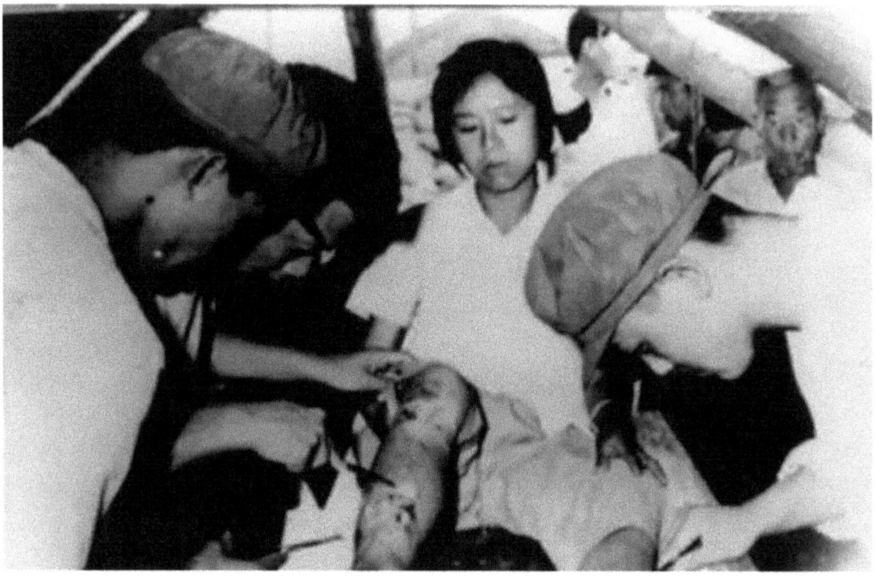

Before the People's Liberation Army arrived, we had to rely on ourselves. We could not sit and wait. Under the organisation and encouragement of Uncle Zhao, people gradually steadied their emotions and regained their confidence. Everyone agreed that we would face the hardship together and not give in. Uncle Zhao made a rough division of responsibilities, covering the care of the injured, medical aid, daily living needs, and security.

Uncle Zhao was tall and thin, sharp minded and alert. He took charge of security. At night he led a few neighbours on patrol, keeping watch to prevent looting or other wrongdoing in the chaos. There were no shoes to wear. They walked barefoot over broken bricks and jagged debris, back and forth across the ruins. Their feet must have been cut and bruised, but they said nothing. Because of their efforts, the households nearby were protected, and people could rest a little easier in the dark.

Uncle Yang was responsible for daily living needs. This was by far the most difficult task. People had gone an entire day without water or food. There was a grain shop nearby, and someone suggested going there to take some grain.

Uncle Yang refused. He said the grain belonged to the shop and could not be taken without permission. Later, someone dug up a few sweet potatoes. We lived on the edge of town, not far from a large sweet potato field. But people felt strongly that we should avoid harming the farmers' interests as much as possible.

Then several local children, Yang Liu, Qi Guoying, Ergou and a few others, discovered a large underground vegetable cellar just north of my home. It was a major storage site belonging to the vegetable company, and some people were already searching there for food. They carried back a basket of potatoes. In that moment, those potatoes meant survival.

Later, water was fetched from the outskirts of the city. Using an aluminium rack from our backyard that had once held chicken feed, we set it over a wood fire and began boiling the potatoes.

The starving people were instantly energised. Hope returned to their faces. As the potatoes were nearly cooked, darkness fell. In the night, several children in their early teens approached us. When we asked them where they were from, we learned that all of them had lost both parents and had become orphans overnight. They lived nearby, in the Huaxin apartment blocks. As the children told the story of their families, everyone present was moved to tears. Kind-hearted Aunt Yang suggested that we must keep them with us and give them the first pot of cooked potatoes to ease their hunger.

From that moment on, we took in the children and shared whatever food and water we had. Later, the People's Liberation Army helped build simple shelters for them and took them back to their original neighbourhood. Only then did we say our tearful goodbyes.

After that, more and more people arrived or passed through. Our place gradually became a relief station. It was not until midnight that each of us finally received a potato that was only half cooked. My father was responsible for medical care. He had once served as a part-time doctor at the factory health station. At a time when the injured were everywhere and medicine was desperately scarce, he announced that there were medicinal herbs in our own courtyard. He had grown a large patch of traditional Chinese herbs himself, lush and green. They were called sanqi and were especially effective for treating external injuries. Remarkably, the herbs had not been buried by the quake. When people heard the news, they came in droves and quickly picked them clean.

My father also taught people how to tend to the injured, how to stop bleeding, and how to apply bandages. Despite his own injuries, he even dug through the rubble with his bare hands, found some medicine in our collapsed house, and used it to treat a child with a high fever, bringing the child back from danger.

The earthquake caused immense loss of life. The devastation was especially severe in the nearby water-machinery workshops where entire families were wiped out. From time to time, children came to my father, asking him to treat their parents' injuries. Their parents had been pulled from the rubble by neighbours, but they would never open their eyes again or speak to their children. In truth, the children had already become orphans. But they chose to believe that their parents were only asleep and would wake up. My father could not bring himself to tell them the cruel truth. They were still so young and still needed care. Yet reality could not be avoided. He could only explain things gently and ask neighbours on all sides to help look after the children. The children still had to go on living. There was still tomorrow.

— Ms Yang

In the early hours of 28 July, the Train No. 129 Direct Express from Beijing to Dalian, carrying more than 1400 passengers, was passing through Guye Station in Tangshan when blue flashes lit up the sky and the train was struck by a violent earthquake. The driver, Zhang Yaowu, made an instant decision to apply the emergency brakes. The train lurched up and down and swayed from side to side. Through calm and alert handling, he gradually brought it to a halt.

When the passengers disembarked, they were stunned by the scene of devastation before them.

Under the coordination of the train crew, over 1400 passengers and staff came together as a united group. Among the passengers were more than a dozen air force officers and soldiers. They took turns carrying large water containers from the train, each capable of holding over one hundred kilograms, hauling bucket after bucket of water from afar. Stoves were set up by the tracks to boil water. This not only met the passengers' needs but also provided a vital water source for nearby disaster victims.

Strangers who had never met before shared what they had brought with them. Sausages, bread, pastries, sweets, everything was passed around. Young people gave up the small cups of water allotted to them so that the elderly, children, and the sick could drink instead.

When the Tangshan Coal Mine district committee and local government learned that Train No. 129 had been forced to stop at Guye Station, they took into account the passengers' living needs. Despite the severe damage in the area, they supplied the train with 2500 kilograms of rice and 100 kilograms of wheat flour. The passengers were deeply moved by the warmth of this shared social bond.

The train staff set up open-air stoves beside the tracks to prepare meals for the passengers. Nearby residents who had themselves suffered losses came upon hearing the news to express their concern and offer comfort.

— Mr Guo

By then, people were unbearably hungry and thirsty and began searching for food and water. Uncle Zhang, strong and energetic, stood up and said to me, “Xiaoqiang, we have to survive. Let’s find something to eat.”

Together, we recovered a small bag of rice and a small cooking pot from the ruins of our house. Using hot water from a thermos flask that had not been smashed, we propped the pot on three bricks amid the rubble and gathered bits of broken firewood, preparing to cook. I asked, “Where can we find matches?” Luckily, Uncle Zhang had gone fishing that night and had been smoking. He still had matches on him. He struck a match, lit the fire, and cooked a pot of thin porridge. We had no chopsticks, so we used small twigs instead. Uncle Zhang and Aunt Zhang, my father, my little sister, and I ate our first meal after the earthquake there on the rubble.

Later, everyone helped us dig my mother and my elder sister out of the debris. The two of them had left us forever. We wrapped their bodies in bed sheets. Two days later, my father and our neighbours dug a large pit on the open ground of a nearby primary school playground and buried my mother and sister there together with more than a dozen relatives from neighbouring families who had also died. Later still, a People’s Liberation Army rescue team removed the bodies for disease prevention. We never learned where they were taken.

That evening, on the very day of the earthquake, people began searching for materials to build makeshift shelters and prepare for the night. My father and I, together with Uncle Zhang, found wooden poles, plastic sheeting, and bed sheets in the rubble and put up a shelter in an open area. My father and Uncle Zhang settled Aunt Zhang, my sister, and me inside the shelter to rest, then went to help other neighbours build their shelters, returning very late.

That night, the city of Tangshan lay in complete darkness. There were no lights at all. Someone dug out a transistor radio from the ruins to listen to the broadcasts. Almost no one slept. We stayed awake until daylight broke the next morning.

— Zhiqiang

Chapter Four

Survivors' Memoir of the 1976 Tangshan Earthquake

Tangshan: A Miracle that Astonished the World

Guo Yan

At 3:42 a.m. on 28 July 1976, a powerful 7.8 magnitude earthquake struck the Fengnan area of Tangshan. With a focal depth of approximately eleven kilometres, the earthquake was shallow, resulting in particularly severe destruction.

This earthquake reduced an industrial city built up over more than a hundred years to ruins. At the time, Tangshan had a population of 1.07 million. Of these, 630,000 people were buried beneath the rubble. The disaster claimed 242,769 lives and left more than 160,000 people seriously injured. Over 7,000 families were completely wiped out, and several hundred thousand households were left without homes. The losses suffered by the people of Tangshan were enormous. A total of 15,886 families were broken apart. There were 8,047 wives who lost their husbands and 7,821 husbands who lost their wives. The earthquake left 3,675 elderly people without children, 4,204 orphans who lost their parents and close relatives, 3,817 people with complete paralysis, and 25,061 people with limb disabilities. Factories ceased operation, mines were shut down, transport came to a halt, and communications were cut off. People were deprived of the most basic necessities of life: food, water, clothing, shelter, and even the ability to reunite with their families.

I.

Before the earthquake, Tangshan was one of North China's most important industrial cities. It was known for coal, steel, and ceramics, and played a key role in the regional and national economy. Although the city accounted for only a tiny share of the country's land area and population, its industrial and agricultural output made up about one percent of the national total. Coal production from the Kailuan Coal Mine alone represented a significant portion of China's overall output.

The earthquake brought this industrial centre to a sudden standstill. Factories and mines were destroyed, transport and communications were cut off, and the city's economic and social systems collapsed overnight. In terms of loss of life, the Tangshan earthquake remains one of the deadliest earthquakes in modern history. The scale of destruction and human suffering was extraordinary, and its impact was felt far beyond the region itself. At the time, the disaster was widely described as one of the most catastrophic earthquakes of the twentieth century. Some foreign observers even believed that Tangshan had ceased to exist.

II.

Yet Tangshan did not vanish. After the earthquake, the people of Tangshan wiped away their tears, buried their loved ones, and turned immediately to rescue and recovery. The enormous destruction caused by the earthquake forced people to find practical solutions to their difficulties. Before the disaster, few had ever built a house themselves or taken part in construction work. After the earthquake, however, large numbers of ordinary residents learned to mix mortar, lay bricks, and repair roofs. They built temporary shelters for their families, constructed makeshift stoves, and gradually became capable, hands-on builders out of sheer necessity.

Food posed another urgent challenge. Many people had nowhere to eat. Setting up temporary communal kitchens proved to be an effective solution. Survivors organised themselves to cook in rotation, eating whatever food was available. If aircraft dropped rice one day, they ate rice. If the next day brought biscuits, they soaked the biscuits in boiling water. When trucks from nearby counties arrived with leeks and flour, people prepared simple leek-filled buns. Life adjusted itself to whatever resources arrived.

Through such daily improvisation and mutual support, the city began to steady itself. At the same time, leadership by example proved to be a crucial factor in restoring even the most fragile sense of order.

The government offices in Tangshan had been reduced to ruins, yet even amid the devastation, responsibility did not disappear. Many senior officials lost their lives in the earthquake. With the original leadership structure severely disrupted, those who survived moved quickly to fill the vacuum. They gathered at Xishankou, where an undamaged public bus was used as a temporary base to establish the Tangshan Earthquake Relief Command Centre. Some of those directing the work still had fresh blood visible beneath the white bandages around their heads. Many had not yet buried family members who had died or arranged care for those who were injured. They set aside personal grief and turned immediately to coordinating rescue efforts and urgent relief.

Through such mutual support, the people of Tangshan revealed a quiet but powerful combination of determination, intelligence, and courage.

III.

The night after the earthquake was bleak beyond measure. Darkness covered the city. A light rain fell as aftershocks continued to shake the ground. People fled their collapsed homes wearing only thin underclothes. Families and neighbours huddled together, calling out names into the darkness, searching for survivors, enduring the long night in silence and exhaustion. Grief seemed suspended between shock and necessity. When morning came, bodies lay along the streets, carefully wrapped in quilts. Beside some of them, voices softly called for parents or grandparents, though no answer would ever come.

In those first chaotic hours, while individuals searched for loved ones and neighbours shared what little they had, the basic framework of the city was slowly being reassembled. It was imperfect and fragile, but it was enough to begin. With the support of the People's Liberation Army and people across the country, Tangshan's residents began rebuilding their city with their own hands. On the tenth day after the earthquake, the first trainload of coal was produced. On the fourteenth day, the Tangshan power plant was reconnected to the grid. On the twenty-eighth day, the first furnace of steel was smelted. Within a year, industrial and mining enterprises had resumed full operation. By the second year, industrial output had surpassed pre-earthquake levels, and agriculture had recovered and begun to grow again. Step by step, signs of life returned to the city.

IV.

Since 2000, the people of Tangshan have pursued a path of reform, openness, and environmentally responsible development. Building on the progress made in restoring the city itself, attention soon turned to coastal development and a strategy of revitalisation through port construction. Jing-Tang Port and Caofeidian Port were successively built, opening more than seventy shipping routes linking Tangshan with countries around the world and providing the city with direct maritime access to the global economy. These developments marked a new stage in the city's comprehensive growth, expanding both its urban functions and its industrial and agricultural production.

Areas once marked by damage and decline have gradually been transformed. What was long described as a "scarred city" has taken on a very different appearance. One of the most striking examples is Nanhu Park, now located at the heart of the city. Covering an area of twenty-eight square kilometres, the site was once a subsidence zone created by more than a century of underground coal mining by the Kailuan Group. Depressions filled with stagnant water, uneven ground, dense weeds, and swarms of mosquitoes had left a visible urban scar. Today, the area has been reshaped into a park of trees, flowers, and open water, rich in greenery and visual harmony. Loved by local residents, it is often described as the city's "green lung" and a natural space for fresh air and relaxation. Each spring, hundreds of thousands of flowering plants come into bloom, drawing visitors from across the region. The park later served as the main site of the 2016 World Horticultural Exposition.

Over nearly five decades of sustained effort, the people of Tangshan rebuilt a modern industrial city. The new city was better planned, more scientifically managed, and more balanced in its development of housing, research, education, culture, and urban and rural life. Tangshan went on to become one of Hebei Province's strongest economies, accounting for roughly one fifth of the province's total economic output. Rather than vanishing, the city re-emerged in a new form and remains a major national base for energy and raw-materials industries.

With a spirit of selflessness, mutual support in hardship, unyielding perseverance, and the courage to move forward, the people of Tangshan refused to yield. The courage shown by the people of Tangshan remains deeply admirable. It was not loud or boastful. It was quiet, determined, and enduring. And it is this spirit that allowed the city not only to survive, but to stand again. In recognition of its remarkable recovery and achievement, Tangshan was awarded the UN-Habitat Scroll of Honour in 1990, acknowledging the city's progress in post-disaster housing and living standards.

The First to Rise from the Rubble

Zhuozi

I was fully conscious throughout the Tangshan earthquake, and I was the first person in Tangshan to stand up from the rubble.

In 1976, my family lived at No. 260 Heping Road, in three west-facing rooms. Behind our house stood a two-storey building belonging to the Tangshan No. 3 Construction Company. On the evening of 27 July, the heat was unbearable. Sweat soaked my clothes. I felt restless at home and could not sleep. Since I was on the night shift, I thought I might as well go to the factory workshop, where it would at least be cooler. I picked up my lunch tin, inside it a bundle of dried noodles, and headed to the Tangshan Knitting Factory.

After arriving, I rested for a while in the stockroom before taking over my shift. I worked as a cloth inspector. My task was to weigh the fabric produced by the weavers, stamp it, and record the worker's number and the weight of the cloth. A little after three in the morning, it was time to prepare food. The factory allowed a simple soup meal to be made between 3 and 4 a.m. for those working through the night.

Usually, the women workers handled the cooking. But that night, on impulse, I said it was time we men took a turn. I asked the shift leader, Yuan Changqing, to join me, and together we carried dozens of lunch tins to the canteen. It was about half past three. I placed the tins along the windowsill and said to the chef, a woman whose smiling face I still remember clearly, "Let us jump the queue. Make ours first so we can eat quickly." She smiled, took our tins, and turned to the stove. Five or six minutes later, our soup was ready. She set the lunch tins down on the other windowsill.

I had just reached out to pick mine up when I heard a tremendous cracking roar, like mountains collapsing, rushing towards us from a distance. Within five or six seconds the ground began to shake. It rolled in waves at astonishing speed.

I managed only two or three steps before the ground threw me off my feet, my body rising and falling with the force beneath me. The violent shaking lasted about a minute before gradually weakening, though smaller shocks continued.

I shouted, "Master Yuan! It's an earthquake! Where are you?"

"I'm here," he answered. Hearing his voice, I felt relief.

The kitchen had collapsed. I called out into the debris, but there was no answer. A chill went through me. I realised that I myself was unhurt—the canteen had been a simple structure roofed with asbestos sheets. I began pulling debris aside and noticed a faint glimmer of light from a nearby window. I cleared away the broken glass and jumped through. I found myself standing on the roof. An unnatural silence hung in the air. There was not a sound. In the distance, smoke rose in several places. A strange thought crossed my mind: Were only Yuan and I still alive in Tangshan? I even wondered how we would ever return to work.

Then the silence shattered. Cries and groans rose from every direction. "Help! Help!" Yuan came out soon after. I said, "Let's rescue the factory leaders first so they can take charge and organise the rescue." The canteen stood next to the duty office. We began digging there, prising up heavy roof slabs. Under one of them we found the security chief, Zhang Yufu. I cleared bricks and stones from around him until half his body was exposed. "I'm all right," he said. "Go and save Secretary Wang."

Secretary Wang was also trapped. We dug him out. The first thing he said when he emerged was, "I'm freezing." I tore down a curtain from the doorway and handed it to him. "Wrap this around yourself." After clearing the duty office, I said, "Let's go to the dormitories. There must be many people trapped there." The dormitories were single-storey buildings west of the main office. We kept digging there for four or five hours. I cannot remember how many people we pulled out.

By then workers from the workshop had arrived. Most had survived because the workshop roof was supported by pitched beams. Only one man, Li Qinglin, had left his post and gone to my workstation, where he was crushed by a collapsing wall. Had I not gone to the canteen, that would have been me.

By daybreak we were exhausted. I lay down in a vegetable plot beside the factory, snapped off handfuls of spinach from the soil, and ate it without even brushing off the dirt. Yuan said quietly, "Who knows what's happened at home." Suddenly I remembered my younger sister. I climbed to my feet and set off towards our house. On the way, I saw a man sitting on the rubble, naked, his leg bleeding. He looked at me pleadingly and said, "Brother, I know your family must be in trouble too. Just carry me to a flat spot and that will be enough." I lifted him by the waist and moved him to a level patch of broken concrete before running on.

Our house was devastated. The two-storey building behind us had crashed down onto our roof. I thought it was hopeless; my sister must surely be gone. Then I heard the faint sound of a radio: "This is China National Radio..." I followed the sound to a narrow crack in the rubble and shouted my sister's name. She replied at once, "Second Brother? Hurry and save me!" She had deliberately switched on the radio to conserve her strength instead of shouting.

Above the gap hung a massive block of brick and concrete weighing more than a ton, ready to fall with the slightest aftershock. I wedged two wooden frames beneath it and began clearing debris by hand. After about an hour and a half, I pulled her out.

By then neighbours from across the street had gathered. We formed another temporary rescue group and pulled many more residents from the ruins. When we finally reached my bedroom, we saw the bed piled high with fallen stones. Everyone said the same thing: if you had been sleeping at home that night, you would have had no chance.

The Heaven's Grace

Xiangju

Every year on the night of 28 July, small clusters of burning joss paper offerings flicker in the streets and alleyways of Tangshan, set out in remembrance of the dead.

Forty years have passed since that season of grief, yet for those of us who survived, the memory feels as close as yesterday. It is not simply that the disaster left scars on my heart. What troubles me still are the hours around 3:42 a.m., the moments before it arrived, the instant it arrived, and the stretch of time that followed. Those fragments have remained a riddle in my mind, gathering like mist and forming questions I have never fully answered, questions I continue to turn over again and again throughout my life.

On the afternoon of 27 July 1976, I accompanied provincial and regional leaders as we drove south from Luan County along the Luan River embankment to inspect flood-control work. As our vehicle passed Macheng in Luan County, the Luan River, which normally flows quietly to the south, seemed to have suddenly grown turbulent. The water level appeared to rise sharply in a short time, and as we continued south, the river kept swelling.

When we returned to the Prefectural Office building, we learned that the upper reaches of the Luan River had experienced several consecutive days of clear weather and that there were no abnormal hydrological reports. A troubling question then settled deeply in our minds: if no rain had fallen in the Chengde region, where had this surging river water come from?

Before leaving work, our driver, Da Zheng, came to see me. "We're heading south again tomorrow," he said, handing me some cash. "Here's the two hundred yuan I drew from the accounts office for petrol. You'd better keep it. I'm not even sure I'll come tomorrow. I don't know what's wrong with me this afternoon. I felt uneasy behind the wheel, as if something wasn't right. If I don't show up by eight tomorrow morning, just find another driver." I laughed and teased him. "You're nearly 190cm and built like iron. You've never even had a headache. How could you not come?" He shook his head and broke into a bright smile before heading downstairs.

I could never have imagined that this strong, indestructible man would be killed at the earthquake only eight hours later. Nor could I have known that his bright smile would be our last farewell.

After he left, I slipped the 200 yuan into a blue plastic-covered notebook, locked it in my fourth-floor desk drawer, and hurried home.

That evening I sat alone at the table, eating freshly steamed vegetable buns. For some reason, I found myself talking to myself. "Poor Xiuzhen," I muttered. "Ten hard years in the countryside, and she's finally made it back to the city. At least I've done my part as her brother-in-law."

Xiuzhen was my wife's younger sister. She had been sent down in 1968. Two months earlier I had helped arrange her return and found her a job at the Tangshan Gear Factory. My wife looked at me in puzzlement. "What's wrong with you tonight? You usually come home without a word, and now you keep going on about Xiuzhen."

I had no answer.

After dinner, Mrs Han lives opposite came over to visit. She held her little daughter in one arm and carried a clay medicine pot in the other.

"Her grandma has gone back to our hometown," she said. "She won't need this for now, so I've brought the medicine pot back."

"It's not a festival or harvest season, nothing special going on," my wife asked in surprise. "Why would she suddenly go back?"

"I don't really know either," Mrs Han said. "This afternoon the child's uncle arrived out of the blue and took her back."

I did not find the old lady's departure especially strange. In any household, small turns of circumstance are always happening. What troubled me, instead, was the medicine pot. In northern custom, a medicine pot may be borrowed, but it is not meant to be returned. To return it is to send sickness and misfortune back with it. Mrs Han was an honest, sensible woman, well aware of such taboos. Why would she do something so clearly considered inauspicious?

The night was stiflingly hot, and neighbours gathered in front of the newspaper office building to catch a bit of air and chat. My wife settled our son to sleep early. I lay in bed reading, as was my habit. I flipped through the books at my bedside yet felt no drowsiness.

I went through every book on my bedside table and still could not fall asleep. At last, I reached for a small booklet entitled Earthquake Knowledge, circulated by the Tangshan Regional Science and Technology Commission. Normally, I would not have given it a second glance. I had been professionally trained, had spent thirteen years issuing weather forecasts, and several more engaged in earthquake monitoring and prediction. The contents were so familiar that I could almost recite them from memory. Yet that night, I found myself reading the booklet with unexpected interest.

In the early hours of 28 July, a time that should have belonged to the deepest sleep, I was still wide awake. I lay there turning the pages of this booklet, utterly absorbed. My wife stirred, glanced at the clock by her pillow, and said, "It's already past three. Why aren't you asleep yet?" I didn't answer. A passage I was reading, quoted from *The Qianlong Annals of Sanhe County*, had completely seized my attention: "On the twenty-eighth day of the seventh month in the eighteenth year of the Kangxi reign, an earthquake struck the capital. Rising from the northwest, sand and dust filled the sky, plunging it into darkness. People were thrown about as if on rolling waves, unable to remain upright. Thunderous sounds followed, sending birds and beasts into panic. The ground shook repeatedly through the night, splitting open the earth and tearing a great fissure beneath Desheng Gate, where water surged up like a spring. The dead and wounded among both officials and civilians were beyond counting; bodies lay piled upon one another, beyond recognition."

I was reading intently when my wife urged me again, "Go to sleep. Don't you have to go out of town tomorrow?" As she spoke, she turned over and held our younger son over the potty. At that very instant, a flash of blue light lit up the entire night sky outside the window. A thunderous roar followed. The ground shuddered violently, and the dust poured down from the ceiling. Frightened out of her wits, my wife loosened her grip. With a dull thud, our son dropped into the potty. I sprang to my feet and shouted, "Earthquake!" The shout was so loud that, had it not been swallowed by the roar of collapsing buildings, it would have jolted everyone in the building awake. My wife cried out in panic, "Go! Open the door!" I yanked our son out of the potty and shouted back, "Don't get off the kang. Run to the window!" I knew that during an earthquake, the sudden air pressure makes doors nearly impossible to open.

After that flash of blue light, cries of terror came from every direction. Then, after a single earth-shattering boom, the entire city fell into a deathlike silence. Moments later, heart-rending screams erupted from all sides, tearing through the dark. Clutching our son, my wife cried out, "I'm not even wearing my trousers!" I snatched the child under my right arm and cursed, "What the hell do you need dignity for now? Even naked, you have to run!" Later, facts would prove this decision to be remarkably wise. In the days that followed, some women who were pulled from the rubble were found still clutching their trousers in death. Sadly, some lost their chance to escape simply because they were trying to get dressed.

Holding our son and dragging my wife along, I had just reached the south-facing window when a tremendous crash sounded behind us. The wall by the door collapsed inward toward the kang. We had narrowly escaped disaster. If I had followed my wife's idea and climbed down to open the door, we would have been smashed to pulp. Amid the roar of collapsing houses and endless screams, I kicked hard at the glass window. With a loud crash it shattered, but the jagged frame left no space to squeeze through. I was trapped, unable to go forward or retreat, sweating in desperation. Then the ground shook violently again. Then the ground convulsed violently again. A crack more than a foot wide split open between the west wall and the window frame. Our family of three squeezed through the gap and stepped onto the outer windowsill. Below the sill was a three-foot-wide passageway, beyond which stood a small shed I had built to store firewood, coal, and odds and ends. Its brick walls and solid roof made it exceptionally sturdy. Just as we were about to jump down from the sill, the shed suddenly collapsed with a hollow roar. Broken bricks and stones piled up level with the windowsill, perfectly forming a path of escape.

I hurried onto the rubble of the collapsed shed. Just as I was about to step forward, the south wall behind us and the window came crashing inward with a heavy thud. I felt a quiet sense of relief. Had we been even one second slower crawling out of the room, we would have been crushed beneath the falling wall. Had we jumped from the windowsill one second earlier, we would have been smashed by the shed outside. I still cannot tell whether this was fate or mere chance. When our family of three reached Xinhua Avenue in front of the building, the wide road lay completely empty. Beneath a concrete utility pole by the roadside, only two figures were moving. It was Li Lianzhong from the newspaper office and his wife. Of the entire newspaper building, they were the first to escape. We were the second.

In truth, from the moment the blue light flashed to the instant I was out of danger, no more than three to five seconds had passed.

We had only just steadied ourselves beneath a concrete telegraph pole on Xinhua Road when another violent aftershock struck. The entire road bucked and heaved under our feet. The pole swayed wildly, tipping one moment, righting itself the next. We dropped to the ground at once and crawled, scrambling on our hands and knees towards the middle of the road. When the aftershock passed, a fine drizzle began to fall. In front of us lay a heap of dark rubble. My mind felt strangely muddled. From somewhere within the debris came the sound of a woman crying out, 'Ouch... Ouch...' I ran towards the voice. In the dim light I saw someone hanging from the south wall of a two-storey building. I shouted to her to jump. The urge to survive left her no room for hesitation. She slid down the wall and I caught her. It was Xiao Lu from the newspaper's printing works. I helped her across to the roadside.

Another figure came hurrying from the east. It was Bi Jin from the Lulong County Committee reporting group. I was bewildered. Lulong County lies over 100 kilometres from Tangshan — what was he doing here at this hour? Later, I learned that he had come to the regional newspaper office two days earlier to revise a manuscript and had been staying in the upstairs guest room with Yi Ren from the Fengnan County reporting group.

Bi Jin is a person who'd like to sleep through the mornings and staying up at night, yet that day he had planned to catch the four o'clock train. He had just reached the No. 116 Geological Building when the main shock struck. Now he had run back to look for Yi Ren.

Yi Ren, unlike him, never sleep in. His revisions were already finished. On the evening of 27 July, he ought to have returned to Fengnan County, just 9 kilometres from Tangshan, but, quite out of character, he insisted on staying one more night. At three o'clock Bi Jin tried to rouse him so they could leave together. Yi Ren grumbled and refused to get up. Just a few extra minutes of sleep cost him his life.

A little after four o'clock, it was fully light. Yesterday there had been factories roaring in every direction, rows of tall buildings standing firm; now, with that blue flash, all of it had become rubble. Beneath the debris, how many were struggling under the weight, how many were groaning in pain...

At that moment, a single desperate thought filled my mind, that someone, anyone, would come quickly to save the people of Tangshan. I was not alone in that hope. Old Li, the caretaker from the newspaper office gatehouse, sat on the ruins and wept aloud, "Lord! Help me, I cannot survive... Chairman Mao, come and save us..." His hoarse cries felt like another blade driven into already wounded hearts. Soon the air was thick with voices calling out, "Father! Mother! My son! My child!".

By around five o'clock, more people had gathered in the street. Those who had escaped with their lives, men and women alike, were mostly half-clothed or naked. Our neighbour Lao Tian had run out completely bare and only later found a bedsheet to wrap around himself. In such peril, saving one's life was all that mattered; shame had no place.

I could not see Mrs Han's family anywhere in the crowd. A chill of foreboding ran through me. Barefoot, I hurried towards the rubble. The moment I took a step, a sharp pain shot through my right leg. There was a cut about an inch long in my calf, clotted with dark red blood, my whole lower leg stained crimson. Limping, I made for the ruins and called out at the spot where the Hans had lived. Only silence answered me.

In my desperation I tried to shift a concrete slab with my hands. It would not move. I looked for something, anything, I could use as a tool, but there was nothing except broken bricks and shattered tiles. I left the rubble in despair. As I walked, I kept thinking, Auntie Han, you left the city yesterday afternoon; was that some turn of fate? And you, Mrs Han, returning the medicine pot last night against your usual habit; did you sense what was coming, settling one last small matter before it arrived?

At about six o'clock, when I returned to Xinhua Road, my wife was holding Zhaozhao and trying to feed him. Her breasts were dry, the nipples shrivelled without a drop of milk. The child cried hoarsely with hunger. Someone had brought vegetables from an orchard. I asked for an eggplant bun no bigger than a walnut. My wife chewed it slowly and fed it to Zhaozhao bit by bit. Only then did he stop crying.

Those who had escaped with their lives stood about with vacant eyes, still in shock, not knowing what to do. When the sun had climbed high, people began to move towards the ruins to search for their buried loved ones. I thought of my father-in-law's family in Majiatun, Lunan District. I thought of my six-year-old eldest son, and of my second elder brother's family in Kaiping. Had all eight of them survived?

I had gone only a dozen steps when Sister Tian from next door came running after me. She was in nothing but a pair of flowered underpants, crying as she ran, "Brother, please, go and save Xiaobao. He's still trapped under the building." The anguish in her voice stopped me where I stood.

When I reached the ruins, Lao Tian was standing there, at a loss. The sight of me stirred him. We began clawing at the debris with our bare hands. Against the wreckage of a two-storey house, it was like ants gnawing at a big bone. My fingernails worn plat. The skin on my fingers split open as I clawed at the bricks. My bare feet were pierced again and again by shards of glass and nails, seven or eight holes in all, and the pain was sharp and constant. I thought of finding a pair of shoes and stepped aside.

Sister Tian caught hold of me at once, "Brother, please don't go."

"I'm not going," I told her. "I'm just looking for something to put on my feet."

I circled the ruins a few times and found a pair of cotton gloves. I pulled them over my feet, tied them on with strips of cloth, and went back to digging with Lao Tian. Brick after brick, we worked in silence. Then my hand brushed against something warm. It was a woman's breast. I called out to Lao Tian. Together we cleared the debris and lifted her out. Blood was seeping from her nose and mouth. She was no longer breathing. Lao Tian said she worked as a cook in the newspaper canteen. I covered her naked body with a bedsheet, and went on digging for Xiaobao.

At about eleven o'clock, we had scraped out a hole barely a foot across. At last, we reached Xiaobao. He was pinned beneath the rubble, unable to move, his body twisted, crying without pause. Sister Tian leapt to her feet on the ruins. "He's alive. Xiaobao is still alive." The relief lasted only a moment. The little opening was ringed with steel bars and broken timbers, and an iron rod lay across the middle. Lao Tian found a wooden pole and pushed it through, trying to prise the space wider. The moment he levered it, Xiaobao screamed in pain. Lao Tian stopped and said, "Whether he lives or dies, we leave it to fate. We have to pull him out through this hole." We took hold of him, one on each side, gripping his neck and shoulders, and pulled with all our strength. As soon as his head reached the opening, it caught on the iron bar. Lao Tian set his jaw. "Pull!"

Through Xiaobao's screams, we clenched our teeth and pulled with all our strength. We got him out, but the iron bar had torn away a strip of his scalp. Blood ran freely down his face. Just then my wife came hurrying over. She was a doctor, and at once she knelt down to bandage him.

After we had pulled Xiaobao out, I set off again towards my father-in-law's home in Majiatun of the southern district. The sun was fierce. I leaned on a stick and limped my way along. At Yonghong Bridge, bodies lay in rows on either side of the road. I could not bear to look for long. The closer I came to the southern district, the worse the destruction. By the time I reached the outskirts of Majiatun, there was no road left at all. I made my way across the rubble, step by step, until I found the house. From a distance I recognised a few familiar figures. Relief surged through me and I hurried forward. My father-in-law's family stared at me for a long moment without speaking.

My brother-in-law told me later that Xiuzhen and my younger sister Xiuzhi had been sleeping together on the kang in the east room. When the earthquake struck, the two of them squeezed down into the hollow beneath the kang. Several roof beams fell and lodged across the edge of the bed, yet neither of them had a single wound on their bodies. Xiuzhi lay quietly in the dark cavity, waiting for the family to dig her out. She was saved. Xiuzhen, however, panicked. She cried out and thrashed, clawing at the air and at the packed earth around her. Ash and soot poured down, filling her nostrils, clogging her mouth. In the end she suffocated there in the darkness. I walked over to where her body lay and lifted the cloth that covered her face. The family had already washed away the ash and grime. Her skin looked pale and clean, her expression calm, without the slightest trace of pain. There even seemed to be the faintest curve at the corner of her mouth. She was only 25. She had just finished ten long years as a sent down youth. Her life was about to begin again, and in a single night the earthquake shattered everything she had imagined. Standing there beside her, I felt a curse rise in my chest against the merciless disaster. At the same time a question pressed on me in silence. The night before, seized by some sudden impulse, I had muttered to myself about wanting to be worthy of Xiuzhen, speaking as if in a dream. And it was Xiuzhen who was taken. Was that strange murmur her final warning to me, or had some hidden nerve inside me sensed the catastrophe that was coming for her, knowing before I knew?

I said goodbye to my father in law's family and set off for Kaiping again, leaning on a crutch and moving as quickly as I could. When I reached the east entrance of Shengli Road, traffic had come to a standstill. Vehicles of every size were crammed together, all loaded with the wounded. The air was thick with shouting and impatience. I pushed my way forward and saw that Jixiang Bridge had split completely in two. The concrete deck had collapsed into a sharp V, the middle sunken deep. The thought of my mother left me no room to hesitate. I slid down the broken slope of the bridge like a child on a playground slide. The river water soaked through my shorts and undershirt at once. Gripping the slabs of concrete with both hands, I dragged myself upward, crawling over the fractured edge until I reached the top and stepped onto the road heading east. The Tanggu Road, once wide and level, was now a field of wounds. Cracks more than a foot wide cut across it in all directions, dark and bottomless. I picked my way forward carefully, testing each step, afraid that one misstep would send me plunging into a chasm. I do not know how long it took to cover those 9 kilometres. By the time I reached Kaiping, dusk had already begun to settle.

I found my second brother's family outside the south gate of Kaiping. He was serving then as the political commissar of the district armed forces department, and he was tied up from dawn to dusk organising rescue work. In the tent there were only my mother, my sister-in-law, and my nephews and nieces. When my mother saw me, she was so happy she did not know what to say. I told her that among the two families in Tangshan, only my wife's younger sister Xiuzhen had not made it through. My mother began to weep. "What a pity," she said through her tears. "Such a fine young woman. Why did it have to be her?" Then she murmured, "Our Meng family must have done some great good in the past. In such a catastrophe, not one of us had been injured or killed."

My sister-in-law told me that the apartment building they had been living in had collapsed only on the western half. The destruction stopped exactly at the wall of their room. If it had given way by even half a room more, Mother would have been crushed. An entire row of apartments had fallen as cleanly as if sliced off with a blade, while the other half stood untouched. During the Tangshan earthquake, she said, nothing like it had been seen. Of course it was a coincidence, an accident of chance. My mother insisted it was the protection of the immortals, that perhaps in some past generation our family had burned enough incense to earn such blessing. I smiled at her. "In the future," I said gently, "don't say that in front of others. It might hurt them to hear it."

After dinner at my second brother's place, the first real meal I had eaten since the earthquake, I set out that very night back to the city. My wife held our son in her arms and squeezed into a large tent the newspaper office had put up. Men and women were crowded together inside, pressed tight in the heat, the air heavy and stale. I could not bring myself to go in. I wandered back and forth outside the tent, my thoughts muddled and unsteady. The ground under my feet felt soft. Only later did I realise I had been stepping on bodies. At the time, I felt nothing.

At daybreak, a harsh blast from loudspeaker broadcast tore through the air, spreading fresh fear among those who had survived. "Another major earthquake is expected today. We urge all residents to leave Tangshan as soon as possible!" People then started to wave down any vehicle they saw, scrambling aboard without asking where it was headed. The scene looked like a defeated army fleeing the battlefield. The broadcast did not frighten me. I only knew that I had to send my wife and our one-year-old son, somewhere safe. Only then could I steady myself and return to the rescue work. I decided to go to Qianxi County. My wife and I had both worked there for years.

When we reached Xibeijin, we happened to meet comrades from the Qianxi county committee who had driven into Tangshan for relief efforts. They told us to wait there. At eleven o'clock, we climbed onto a truck bound for Qianxi. The roads were jammed, and the 25 kilometres to Fengrun took more than three hours. It was past four in the afternoon when we finally arrived at the town of the county. Tents had already been set up in the courtyard of the guesthouse, and many of the injured were settled there. After arranging a place for my wife and son, I caught another vehicle and headed straight back to Tangshan.

On the night of the 29th, I wandered in the north courtyard of the newspaper office until dawn once again. On the morning of the 30th, I was about to head for the prefectural committee offices when I heard someone shouting my name at the top of his lungs. I looked up. It was Wang Wentian, the Party branch secretary of Caigezhuang in Laoting County. Behind him stood dozens of sturdy young men. I felt a surge of emotion and stumbled toward them.

He told me that as soon as the villagers heard that Tangshan had been struck by a major earthquake, they had been beside themselves with worry, hanging in suspense over our safety. He had gathered several dozen core militiamen. Carrying crowbars and sledgehammers on their shoulders, they had walked more than 50 kilometres to Tangshan to rescue those of us who had once worked in Caigezhuang. His words caught in my throat. I could not say a single sentence.

When he learned that my whole family was safe, he directed the men to start digging through the rubble where our things had been buried. The young men swung shovels and raised pickaxes. Before long they had uncovered everything I owned, two wooden trunks filled with clothes. When they turned to leave, I did not offer a single word of courtesy. Words would have sounded thin. I took each calloused hand in both of mine, one after another, feeling the roughness in my palms. I stood there watching their figures recede into the distance, growing smaller with each step, until they blurred before my eyes. The tears came again, warm and unrestrained, and I let them fall.

From that point on, I felt unburdened. There was nothing left tying me down, nothing left to cling to. I threw myself into the relief work alongside the comrades from the prefectural committee. Around 3rd August, the finance section began settling public accounts. When they asked about the two hundred yuan the driver, Old Zheng, had advanced for fuel before the earthquake, I told them the money was still in my desk on the fourth floor, buried under the rubble.

Later, someone pulled me aside and spoke in a low voice. "There are people questioning where those two hundred yuan went." I felt my throat tighten. "Heaven and earth can bear witness," I said. "In a disaster like this, would I really stoop to profiting from two hundred yuan? Would it be worth it?!"

Not long after, the Tangshan relief headquarters decided to relocate to Luan County. I worked there for more than three months and returned to Tangshan in early November. The moment I stepped back into the offices, Sister Xu told me the money had been found. It had indeed been tucked inside a notebook. She placed the blue plastic covered notebook, dug out from the ruins, into my hands.

Happiness Buried in the Rubble

Guangsheng

27 July, 1976. The heat that night was thick and airless. I had fallen asleep, though I do not know for how long, when a tremendous crash jolted me awake. I had just turned over when I saw the window in front of me, the doorframe, and a section of cracked wall collapse with a thunderous roar. Instinct took over. I bent into a half kneeling crouch and threw my arms over my head. In the next instant, the falling debris swallowed me whole. I struggled with all my strength, twisting my body, but I could not move at all.

"What's happening? Is it an atomic bomb?" I heard my father shout somewhere nearby, his voice distant and unreal. "It's not an atomic bomb. It's an earthquake, an earthquake!" I forced the words out, though I could barely breathe. In the moment the house came down, I heard my mother's voice from her throat, low and raw, "Ah... ah... ah..." The sound faded after a short while. Later I learned that a rolling stone had struck her on the head. The bridge of her nose had been crushed flat.

When my two nephews and our neighbor Old Xu finally dug me out, it was already full daylight. A fine rain was falling.

As I looked around, I saw that within several kilometers every building had been flattened. Only farther off, three or four kilometres away in the direction of the Tangshan Locomotive and Rolling Stock Works (CRRC Tangshan Co., Ltd.), an elderly woman knelt in the rubble, clutching a body and crying out without restraint. Near me, besides my mother who had been killed, lay my elder brother's son and daughter. My brother himself had been crushed, his breastbone and collarbones broken in several places. He could no longer speak and lay on the debris groaning softly. People all around stood, crouched, or sat among the ruins like wooden figures, staring straight ahead without expression.

I squatted there numbly, my chin resting in my hands. My mind was completely blank. Three or four days later, I learned that my second brother was still alive. He had been injured and sent elsewhere to recover. His wife and their two children were gone. About a week passed before my two nephews and I managed to rig up a crude shelter with sheets of plastic. One after another, we dug out the bodies of my mother, my niece, and my nephew, carrying them to a spot beneath three large locust trees nearby and burying them there. After that, we began to talk about going to retrieve the bodies of my second brother's wife and children. Five of us went that day to the ruins of the staff housing at Tangshan No. 4 Middle School, carrying worn shovels and pickaxes. Teacher Li Suzhen, who had lived next door to my brother, told us that when she crawled out after the quake, she could still hear the younger child wailing beneath the rubble, but he was buried too deep for her to reach him. They had first pulled out the elder boy from a shallower spot. He was still alive then, though his pupils were already dilated and his forehead bore a wound from a falling stone. Teacher Li gave him a few mouthfuls of water. Not long after, he died. When it rained, someone had covered his body with a sheet of plastic. During the clearing of corpses in the following days, he was taken away. It took us nearly half a day to open up the deep layers of debris. When at last we reached them, the sight before us stopped me cold. My second sister-in-law and the younger boy were wrapped together in a mosquito net. Both bodies had badly decomposed and gave off a heavy stench. Their skin had turned dark and purplish. It did not look like death from falling stone. It looked like they had been smothered alive beneath the rubble. The younger boy lay on his back. He had torn himself free from his mother by more than a meter, his body still held in a posture of straining forward, as if he had been trying to crawl out. My second sister-in-law lay face down, her body twisted toward the younger boy. One arm was stretched straight out in the direction he had been crawling.

The five of us gathered the scattered bones of my second sister-in-law and the younger boy into a military body bag. On the eastern side of the No. 4 Middle School athletic field, we scraped out a pit nearby and buried them there. I found a piece of wood and, with a stick of chalk, wrote on it: "The grave of Second Sister-in-Law Li Zhiming and her sons Liu Ran and Liu Zheng." I set it upright at the head of the mound. Only days earlier, that same field had been filled with their voices and laughter. Now it was their grave. I knelt in front of the mound and wept without restraint. I bowed again and again, striking my forehead hard against the ground. When the skin split and blood seeped out, I did not feel it.

Forty full years have passed. The sound of those blows still trembles somewhere inside me. It has never fallen silent. Whenever I think of the cruel truth that my second sister-in-law and the boy suffocated there alive, a pain rises in me that I cannot name: you weren't crippled, you could move, you could crawl. Why didn't you go straight to the school dormitory and dig them out?

Afterward, my second brother told me what had happened to him. He had been pulled from the duty room at No. 8 Middle School with both collarbones broken. The leaders ordered him transferred at once for treatment. He lay on the athletic field of the school and refused to leave. His reason was simple. He wanted definite news from home. He wanted to see with his own eyes whether his wife and two sons were alive or dead. He believed that if his wife were still living, she would crawl if she had to and come to the school to find him. He waited for days. What came instead was word from the No. 4 Middle School that his wife and both boys had been killed. Only then, in a grief beyond bearing, did he allow others to help him onto a helicopter bound for Cangzhou for recovery.

More than a month later, he and the other lightly injured were sent back to Tangshan. When the large coach carrying them drove into the city, past endless rubble and makeshift shelters pitched at random, the escorts said with concern, "Please tell us roughly where your homes are. We will take each of you there." One by one the wounded identified their neighbourhoods and got off the bus. Only my brother remained seated.

"Where is your home?" someone asked him gently.

He could hold back no longer. He broke down, his voice torn by sobs. "What home do I have now? The earthquake took them all. I'm the only one left."

No one present could keep from being moved. After repeated urging, he was finally persuaded to go, for the time being, to stay near Liutun in Lunan District with our father.

My second brother had taught Chinese at middle school his entire life. He was 42 the year of the earthquake. After such a blow, he could not free himself from it for a long time. For several years after the earthquake, he wore a wide black armband without fail. In Tangshan it was common to see people in mourning with black armband, but to keep it on year after year, never taking it off, was something else. In the prefabricated room where he lived after the disaster, he kept a framed photograph on the wall, edged in black. It showed him with his wife and the two boys. On every memorial day, he set out small offerings before their portraits. At night, he would go alone to the nearby crossroads and wander there in silence.

He loved browsing in bookstores. In those years, whenever he bought a book he liked, he would write on the title page, "Dedicated to my beloved wife Zhiming, and to my sons." When he sat talking with friends, it never took more than a few sentences before the conversation turned back to his family, to his wife and the boys. "I'm becoming like Xianglin Sao (one of those people who never stop telling the same sorrowful story)," he once said of himself, half bitterly.

Not long after the earthquake, many of those who had lost a spouse remarried. My brother did not. It was seven or eight years later that, with the urging of others, he reluctantly formed a new family. He could not let go of the past. In the end that marriage, too, fell apart. He is 72 now. Two years ago, he retired from the university lectern. Today he lives alone in Tangshan.

I Narrowly Survived the Great Earthquake

Zixing

In 1976, I had just turned ten. The night of July 27 was unusually dark. In the countryside back then, people ate their summer meals in the courtyard, where the air moved a little and the heat inside the house could be avoided. That evening we were eating in the yard. My second brother, who drove a horse cart at the commune textile mill, said he did not feel like going back to sleep at the factory. My mother would not agree. He had not slept at home in years, she said. There was not even a proper place for him to lie down. Besides, he had to be up before dawn to take the cart out, which meant getting to the mill early to feed the animals. If he stayed at home, someone would have to wake him in the middle of the night so he could walk back. After some urging, he finished his bowl of rice and left reluctantly. Before long, though, he returned. He said the road was too dark. When he reached the stretch by the reed marsh halfway there, a chill rose in his chest. He could not bring himself to go on.

So it was that, for the first time in five years since leaving home to work at seventeen, my brother spent the night under our own roof.

That night was unusually stifling. As my father later recalled, he felt so hot that he could not fall asleep. He lay there fanning himself with a palm-leaf fan for most of the night. Eventually, he got up and went to the east room.

At that time, our house consisted of three tiled rooms, with a central hall in the middle and living rooms on either side. Our family slept in the west room, while the east room had never been renovated. It was used to store odds and ends and was not lived in, so it was relatively cooler. After going into the east room, my father found a flat spot and lay down. He later remembered that not long after he lay down, the earthquake struck.

When the earthquake came, I was jerked awake from sleep. I remember that moment clearly. I slid off the kang and dropped to the ground, then ran toward the main room. The west room had two doors that opened outward. The southern door was already open. My mother had flung it open as she pulled my fourth brother toward the hall. The northern door was still closed. Just as I reached for it and pulled it open, the house came down. The roof collapsed and I was buried at once beneath tiles and brick. I lay face down under the rubble, unable to move. My body would not answer me, yet my mind was clear. I do not know how long it was before I heard my father calling. In the darkness he was shouting our names, his voice hoarse and torn. I answered him, but my own voice sounded faint and far away.

The main beam in the west room had split in two. Its fall brought down both the west room and the main room. The east room remained standing, and that is why my father survived. When he finally crawled out from there, he found only ruins before him. As he called out to us, he saw my second brother standing by the corner of a wall. Later my brother said that when the shaking began, he did not know how he had ended up outside. He only realised he was standing by the wall when he looked up and saw the stars. It was my father's shouting that told him an earthquake had struck. My mother and fourth brother were pinned only from the waist down. They could still speak. I alone was buried all the way at the bottom. My father kept talking to me the whole time. With his left hand he reached into the rubble and felt for my head so that falling bricks would not strike me. With his right hand he pulled tiles and stones away. My second brother used both hands to clear the debris from below. I remember little else from that time. Only one sentence from my father has stayed with me. I think my fourth brother was crying for him to come first. I heard my father say, "Stop shouting. Your third brother is buried the deepest. We'll get him out first." I do not know how long they dug. At last, between my father and my brother, the rubble loosened and I was pulled free. Not a single piece of skin on my body was broken.

At first light, many of our neighbors came running. Together they dug my mother and my fourth brother out from the wreckage. My fourth brother was unhurt. My mother's left hip had been crushed and broken. Three days later she was taken to Fengrun Railway Station and lifted onto a train. In the beginning we had no idea where she had been sent. Each day we worried about her, not knowing what hospital she was in or who was caring for her. Then, sometime later, a letter arrived from a hospital in Zaozhuang, Shandong. It had been written by a doctor on her behalf. Only then did we learn where she was. It is a pity that the letter was later misplaced. My mother could not read or write, and she had not remembered the hospital's name, its address, or the names of the doctors and nurses who treated her. That has remained a quiet regret in our family ever since.

On 8 September of that year, before she had fully recovered but insisting on coming home, my mother was put on a train. On September 9, somewhere along the journey, news reached the train that Chairman Mao had passed away. I think she arrived home on 10 September. I was not in the house at the time. Someone came running to tell me that mother was back. I hurried home at once.

She told us that the doctors and nurses in Zaozhuang had taken very good care of her. She should not have been discharged so soon, she said, but she had longed for home, and the hospital had reluctantly agreed to let her go. For a long time afterward, as she mended, she would often speak of those weeks in Shandong. Each time she did, she would say that it was thanks to the doctors and nurses there, who had treated the injured as if they were their own family.

In the days after the disaster, there was little weeping. Those who had lost family, those whose homes had collapsed, most of them did not cry. It was not that people had grown numb. It was simply that in the face of such devastation, there were things that pressed harder than sorrow. There were lives to save, shelters to build against aftershocks, food to find. There was no time to sit with grief.

I think of my second brother digging me out. A nail had pierced straight through his toe. Blood ran down, and he did not even notice. Only after we had all been pulled from the rubble did he feel the pain and see the blood soaking through.

Later, a medical team from Shanghai came to our village to treat the wounded. I can still see them in white coats, going from house to house. Sometimes they carried a large bucket of herbal broth brewed from wild greens, ladling it out bowl by bowl for people to drink.

After a while, our family received government relief funds. The following year we rebuilt the house. My mother's condition improved little by little. At first, she could only get out of bed. Then she began to move about the yard. In the end, she recovered completely.

Memory Never Fades

Chunlong

28 July 1976 is a date no one in Tangshan can ever forget. What we lived through in that earthquake, which stunned the nation and the world, has long since settled in our minds and is impossible to forget.

I still remember the early hours of 28 July. People were fast asleep when a sudden gale seemed to rise from nowhere. A strange blue light flickered. It felt as if the sky and earth were splitting apart and the earth turning over. I was jolted awake and sat up abruptly. Instinctively, I scooped up my daughter, who had just turned one. I turned to wake my husband. I caught only a glimpse of him before he vanished from sight. In the next instant, everything fell into darkness, and our whole family was buried in the rubble.

In the darkness, I could still feel the earth trembling. Terror overwhelmed me. Holding my daughter tightly, I curled myself around her, frozen and helpless. When the shaking finally stopped, a small opening appeared above and to our right. A thin beam of light shone through. I was suddenly filled with hope. I quickly clawed the opening wider and crawled out. By sheer luck, my daughter and I made it out of the tangled wreckage with only minor scratches.

Standing on the rubble, clutching my child, I trembled uncontrollably and shouted for help. "Help us! Her dad is still buried in there!" Hearing my cries, my mother-in-law rushed over from the neighbouring house. She took the child from my arms and urged me, "Go dig for his father!" I stared at the ruins, my legs shaking, not knowing where to begin. At that critical moment, my two sisters-in-law ran over. Together they lifted a massive slab of the collapsed roof. I steadied it from the side as they climbed inside, calling out my husband's name again and again. After several minutes, a faint response came from within. One of my sisters-in-law cried out in relief, "He's here!" My husband was completely buried beneath bricks and earth. His exact position was impossible to see. My two sisters-in-law dug frantically with their bare hands. When their hands were already torn and bleeding, they finally uncovered him. When he was pulled out, his face was dark and badly swollen, his body burning with heat. He could not stand on his own. With everyone's help, he was dragged from the rubble, his life narrowly saved.

But our fourth younger sister, separated from us by only one wall, was not so fortunate. When the earthquake struck, a massive section of wall crashed down and forced her into the hollow beneath the kang. By the time they pulled half of her body out, she was no longer breathing. She left us there, without a word.

Even now, whenever my husband thinks of that night, a dull ache rises again. After so many years, the pain still lingers in him. He once said to me in a low voice, "If they hadn't tried to save me first, maybe they could have dug her out sooner. Maybe... maybe it wouldn't have turned out that way."

I remember the night of 27 July. The heat hung low and heavy. We were all sitting in the courtyard, talking to pass the time, when someone knocked at the gate. It was my husband's colleague, Xu Weilian. He had come by to see "Second Brother Lin." My husband's surname is Lin, and at home he is the second son, so everyone at work called him that. That night the two of them talked until after one in the morning. Only then did Xu rise to leave. Two hours later, the earthquake came. That strong young man, not yet married, was gone with it. He was 26. His parents lost their only son. It was more than they could bear.

I also remember that same night at my parents' home. The courtyard there was bright with lights. Laughter drifted out from time to time. Three families, ten people in all, had gathered to celebrate the successful match between two older unmarried young people. The celebration carried on deep into the night. At 3:42 in the morning, five of the ten people who had been laughing together were dead. Among them were my father, my elder sister, and her daughter, as well as the two young people whose meeting they had been celebrating. They left us there, all at once...

The earthquake that stunned the nation and the world happened nearly fifty years ago, yet it still feels as though it took place only yesterday. Some scenes remain so clear that they return without effort. When we crawled out of the rubble, the first thing we heard was crying. Then the shouting. Everywhere we looked there were ruins. In front of almost every home lay the bodies of those who had died. Aftershocks came again and again. A fine rain began to fall. We had nothing to eat, nothing to drink, and stood there letting the rain soak us. Fear and grief hung over everyone.

I remember how blank my mind felt. I could not think beyond the next hour. I did not know whether we would still be alive the following day. In those most helpless moments, the People's Liberation Army arrived. They pulled people from the ruins. They cleared debris. They helped us put up makeshift shelters. When they left, they gave us a military shovel so that we could continue digging out what little remained of our belongings. That shovel stayed with us for a long time. Whenever I see it even now, I think of those days when soldiers and civilians worked side by side. The memory rises in me like a quiet warmth.

Help also came from every direction. Supplies poured in from across the country. Medical teams arrived one after another. Construction crews followed, building new apartment blocks where rubble had been. Bit by bit, Tangshan said goodbye to the temporary shacks and stood up again. We returned to work. We rebuilt our lives.

Today we live in bright, spacious homes. Our family enjoys a peaceful life. I often think that the children who come after us should remember this: disasters may be merciless, but people are not. Without the selfless help that came from all over the country, without our own resolve to rebuild, there would be no life like this now.

My Earthquake Memories

Chen He

Forty years have passed since the Tangshan earthquake. Whenever I pick up a pen and try to write about that instant in the early hours of 28 July 1976, my thoughts fall into knots. The fear and grief from that night rise together, and I find myself unable to go on. Over the years I have tried again and again, more times than I can count. Each time I manage only a few lines before setting the pen down.

I was there. I saw with my own eyes how disaster descended in a single moment; how helpless people looked beneath the collapsing sky. I also saw something else. In the days that followed, when everything seemed lost, the people of Tangshan did not give in. They saved themselves. They helped one another. They held fast together through those bitter months. Unless someone has lived through it, I do not think it is possible to understand the feelings we carry toward that part of our history. They are tangled and hard to name, even now.

In 1976 my husband and I were both 35. We had two children, a ten-year-old boy and an eight-year-old girl. We lived at No. 3 Xiaobei Hou Street, Gou East, in the Lunan District. My mother-in-law, already past sixty, happened to be staying in Beijing with my elder sister, helping care for a child there. She did not return to Tangshan until after our house had been rebuilt. That summer, to escape the heat, the four of us moved from the side room into the main room, which was higher and cooler. It was a small decision made without much thought. Later I realised that it was this small shift that spared our lives.

The night of the twenty seventh was unbearably hot. The air felt heavy, pressed low. Even the young men in the courtyard said it was hard to breathe. Indoors or outside, the heat wrapped around us like a thick quilt, leaving everyone damp with sweat. After dinner, the children were playing in the courtyard. Under the telephone pole a few men sat playing cards. The rest of the neighbours pulled their chairs outside, talking and trying to catch a bit of air. The chickens in each household behaved as usual. There was nothing to warn us. If not for what followed, it would have been an ordinary, peaceful summer night. As the sky darkened, the adults drifted back inside. The children had long since fallen asleep. When I went in, I saw that they were lying with their heads toward the window. Normally we made them sleep with their feet toward the window to keep them from catching a draught. It was already past midnight. I did not wake them to change their position. I lay down the same way, with my head toward the window. Looking back,

I sometimes think that without this chain of small accidents, we might not have survived. In the heavy heat that refused to lift, we had no sense that a calamity beyond imagining was about to arrive. We drifted into sleep, unaware that fate was already moving toward us in the dark.

It was already very late when we lay down, and the heat would not lift. I was exhausted, yet I never quite fell asleep. I was half-asleep when I suddenly heard a strange sound, something like a distant r*umble... rumble*. It seemed to come from far away, yet at the same time from deep beneath the ground, low and frightening. *Rumble... rumble...*

I jolted upright instinctively. The Kang beneath me bucked violently. My husband and I jumped to the floor at the same moment. He shouted, "Earthquake!" and reached for the light. The bulb flared white like a flash of lightning across the dark room, then went out with a sharp snap. In that split second, I saw the walls quivering like sheets of paper. I screamed. At the same time, I grabbed the legs of our two children and dragged them off the bed. Before I could pull them under me, the house collapsed with a deafening crash.

I was frighteningly clear-headed. I reached out but could not find my kids. Thick smoke and choking dust filled the air, swirling like a dense fog. I could see nothing around me, and I could not see my husband. My heart sank. The sudden disaster left my mind completely blank. I thought: "Where am I? Where is our house?" Just then, another explosion thundered through the darkness. Red flames tore through the haze. A shop not far from our home was on fire, and the fire ignited the oil drums stored in the shop. By the firelight, I forced myself to calm down. My husband and children were nowhere in sight. As far as I could see, every house had collapsed.

Where are they? I began calling my husband's name. At first there was no answer. After calling several times, I finally heard his weak reply: "I'm by the door." I steadied myself and realised that I was standing on what remained of the roof. My husband and the children were buried beneath the rubble. The roof under my feet, about half a foot thick, had already split into four pieces. In the darkness, I caught sight of a man's silhouette to the east.

"Please help me!" I shouted.

"Don't come over," he called back. "I'm not wearing any clothes."

"I don't care," I shouted. "Just help me save them!"

No matter how much I pleaded, he would not come. In desperation, I picked up a rafter as thick as an arm from the rubble and thrust it beneath the slab pinning my husband down. I leveraged it with all my strength. The slab did not move at all. Instead, the timber snapped in two. Long after the earthquake, I realised that I could never have had such strength under normal circumstances. It was the extremity of the moment that forced it out of me. When the rafter broke, I pulled out another beam like a madwoman and wedged it beneath the roof slab again. I shouted once or twice more, hoping someone might help us, but there was only dead silence.

"Is anyone there?" my husband called.

"There is someone," I replied, "but he won't come no matter what I say."

"Then stop shouting," he said. "Save your strength. You'll need it to dig us out."

I stopped calling. Under his instructions, I pried the roof up little by little, propping it higher each time with bricks. When it was raised enough, I finally saw my husband and our son. My son's upper body was exposed, his lower body still buried, but he was breathing. I lunged forward and tried to pull him free, but it was useless. "Mum," he said, "don't worry about me. I can still breathe. Get Dad out first." At that moment, a violent aftershock struck. The ground rolled beneath me like a small boat caught in heavy seas. Barely able to keep my footing, I staggered to my husband and saw that a stone slab lay across his neck and shoulders, as large as a school desk. There was no time to think. I began clawing at the earth, fragments, and stalks of sorghum used in the roofing with my bare hands. At first, I heard my husband urging me urgently, "Hurry! Hurry!" After those two cries, his voice began to fade. Panicked, I reached blindly into a gap and grabbed hold of his hair. Summoning all my strength, I pulled hard and dragged his entire body free. He gasped for air. "I almost suffocated." Only then did I realise he was soaked through, sweat and dust mixed into mud all over him. Seeing that he was safe for the moment, I rushed back to our son. By then, my son had freed one arm with the other and had spotted his little sister beneath the rubble. He cried out, "Big Bubble!" I ran over and saw my daughter. A huge blister on the back of her neck rose and fell as she struggled to breathe. A deadly slab of stone pressed down on her head. I was almost despairing, but I tried the same method again. I thrust my hands under the slab and pulled with all my might. She came free.

My daughter's nose looked crushed flat. Tiny red specks covered her face. She was soaked through with sweat, her whole body smeared with dust until she was like a small figure of mud. On her shoulder there was a gash shaped like an infant's mouth. When I looked closer, I understood that her nose had not been crushed. She had not been able to breathe. Her head had swollen from lack of air. Her entire face was so puffed that I could no longer see the shape of her nose. Her eyes were vacant. She did not speak. She could not even sit on her own. I pinned her against my side with one arm and began clearing debris from my husband with the other. Several of his ribs were broken. He could not use any strength. I told my son to hold his sister upright. Then I slipped both hands under my husband's arms and dragged him out in one pull.

A shard of stone had torn open his calf, leaving a gash nearly a foot long. Blood streamed down his leg. We found a pair of trousers, tore them into strips, and tied the wound as tightly as we could. After binding his leg, I struggled back to free my son from the rubble. By some mercy, he had only a few scrapes. Then I gathered my daughter into my arms and began rubbing her limbs, pressing her chest, urging her to breathe. Slowly, very slowly, her breathing steadied. It was then that I felt something swinging near my left eye each time I moved. I reached up and tore it away. Only when I looked down did I see that it was a flap of flesh from my forehead. I had not felt any pain. Just then I heard the old uncle from our courtyard crying out for help. I ran toward his voice. We pulled him out first, then helped rescue his wife, his son in law, and his granddaughter. There were no proper tools. We clawed at dirt and at the straw packed into the roof with our bare hands. We were snatching people back from death. Half a minute's delay could mean someone would suffocate beneath the ruins.

Everyone kept digging, rescuing themselves and others. More and more people were pulled from the rubble. Those who were slightly injured, or not injured at all, did not rest for even a moment. They joined in. Hands worked without pause. As the sky slowly brightened, rain began to fall. Only then did I begin to feel the pain everywhere in my body. My fingers were swollen like carrots. I could not close them. Mud was packed beneath every nail, and when rainwater touched them, the pain was sharp and unbearable. I touched my face and found my hand covered in blood. My left brow bone had been smashed. The muscle above the eye was torn.

We did not know whether anyone would come to save us. Someone managed to pull daily necessities from the rubble and found a small transistor radio. I said, "Quick, tune to the central broadcast. If Beijing hasn't been hit, we'll have a better chance." People gathered around the radio. When they heard that the Central People's Broadcasting Station was still transmitting as usual, it felt as if hope itself had spoken. We stood a little steadier.

By then I was exhausted and in pain. I thought of finding even half a quilt from the side room so I could lie down for a moment. But the damage there was beyond anything I had imagined. Several unused quilts on the kang had been driven straight into the hollow beneath by the collapsed roof. There was no way to reach them. That meant that if we had slept there that night, we likely would not have survived.

After the earthquake, our greatest difficulty was the lack of basic supplies.

Adults could endure for a time. The elderly and children could not. By the afternoon of the twenty eighth, we had gone more than twelve hours without water. My children were small. My husband had fractures all over his body and could not move. I myself had a fractured pubic bone, but I still had to force myself toward the river for water. Otherwise, we would face heatstroke and dehydration. I took a few steps, then stopped to fight the pain. In the end, I sat beneath a roadside tree and closed my eyes, exhausted. When I tried to stand again, my body would not lift.

A woman passed by. "Please, little sister, help me up," I called.

She let out a scream and ran away.

Perhaps she thought I was already dead. My face was swollen from lack of air and smeared with blood and dirt. Later two passersby helped me to my feet. When they heard that I was heading to the river, they told me it was impossible. The ground had split and shifted, forming a steep drop. Even the uninjured struggled to cross. I would not make it.

I had no choice but to turn back, limping home empty handed. As I approached the railway near our house, I saw my daughter picking up discarded tomato stems someone had thrown away after eating. I called out to her. She quickly hid her hands behind her back, afraid I would see.

My heart tightened. "Throw it away," I said softly. "It's dirty. You'll get sick. When the uncles and aunties carrying baskets come by, ask them for something to eat. The vegetable plots are across the river. Your father and I are both injured. We can't make it over there." Later she managed to obtain some tomatoes, green peppers, and cucumbers, and shared them with the other children, while the adults went hungry.

I had thought that the second day after the earthquake would pass the same way, in hunger and thirst. I did not expect that Director Zhang from our factory would find us. He had been on his way to look for the secretary. As he moved through the area where many former steel mill workers lived, he went from place to place, looking for any of his employees who might still be there. That was how he came across us.

He told me to go to the factory right away. There was clean water in the bathhouse. Two large pools had been filled the night before, and the earthquake had struck before the workers could use them. The moment I heard this; I hurried off to find a flat cart that had been used to move the bodies of the dead. I loaded my husband and the children onto it, along with the son in law of the uncle from our courtyard whom we had rescued. The old couple had only minor injuries and could walk, helping to push the cart. Even with the cart, my leg throbbed without relief. I could only move a few steps at a time, stopping often. When we reached the former Qixin Club, we were fortunate to find a medical station set up by the East Sea Fleet. We joined the line to have our external wounds treated. An elderly military doctor cleaned our injuries one by one, asking about our situation as he worked. When he learned that we had gone two days without food or water, he immediately had someone bring two large bottles of glucose solution. That sweet liquid was like life itself. Both families drank it. The coolness flowed down my throat and seemed to steady my heart. We gathered what strength we could and continued toward the factory.

The road felt endless before we finally reached Huaxin Bridge. I told myself we must have some luck left, because a group of militia were maintaining order there. We happened upon a man carrying a case of beer across the bridge. The militia stopped him and allowed him to take only two bottles. We were weak from thirst and hunger. I stepped forward and asked if we might be given the two bottles. When they saw how many of us were crowded onto that cart, they handed two bottles to us. The adults and children shared the beer. The children did not get drunk. Even I, who am allergic to alcohol, had no reaction at all.

In the days after we reached the old rolling mill factory, we no longer had to worry about food. The factory still had stores of grain and vegetables, even some biscuits. There were no proper containers, so we used washbasins and safety helmets to hold the food. At that moment, it tasted better than any feast I had ever known. The central government moved quickly to organise relief, even though the shadow of the Cultural Revolution had not yet fully lifted. Troops and medical teams arrived from all over the country, batch after batch. Each time I saw another group of strangers step down ready to help, I felt our chances grow a little stronger. Every day seemed to bring something that felt like a small miracle. The railway lines were restored with astonishing speed, reopening the way out of the city. Seriously injured people, like my husband, were transferred in orderly groups. My elder sister and my mother, already in her sixties, were pulled alive from the ruins of their apartment building after being trapped for three days and three nights. The bodies of the dead were handled quickly and with care, and no epidemic took hold. Behind each of those days stood countless soldiers, doctors, nurses, and relief workers who did not rest. It was they who helped the people of Tangshan stand against the disaster, racing against time itself.

Reconstruction began almost immediately after the disaster. Summer would soon give way to autumn, and winter would not be far behind. If we were to survive the cold, we needed a shelter to live in. I took my two children back to the ruins of our courtyard and started planning how to build a home again. By chance, my brother-in-law had been sent on relief duty. He came with a convoy of trucks from Beijing, following roads warped and twisted by the earthquake until he found us. He wanted to take both the children and me back to Beijing.

I refused.

"Take the children," I told him. "I have to stay. Everything we own is here. If I leave, there will be nothing left when we return. Once the children are safe, I can manage the rest." He tried to persuade me, but in the end, he had no choice. He took the children and drove back to Beijing.

Not long after, I found myself sitting alone on the rubble, guarding what was left of our "home." A few soldiers from the People's Liberation Army noticed me. Seeing that I was by myself and still injured, they hurried to build a small makeshift shelter for me. They helped dig out our daily necessities, along with the sewing machine and bicycle that had not been destroyed. The salvaged belongings, food, and household items were piled high across what was left of the roof. I did not even have clean water to offer the soldiers. So many feelings welled up inside me that I was on the verge of tears. I could only bow to them again and again, trying to steady the tears that threatened to spill over.

After a period of time, the old rolling mill factory provided building materials for those of us who worked there. Tangshan Iron and Steel Group sent men to help me put up a simple dwelling. At last, I had a place where I could get through the winter. The neighbourhood committee distributed vegetables and meat at regular intervals according to headcount. Drinking water was delivered by the soldiers. My leg, though never operated on, slowly healed on its own. During that time our wages continued to be paid as usual. I received a letter from my husband in Laiwu, Shandong. He wrote that his injuries were nearly healed. The hospital took care of treatment, meals, and clothing. Everything was fine.

The disaster destroyed our home and showed us how fragile life can be. Yet what has stayed with me even more are the people who laboured in the rescue and rebuilding. Their efforts taught me something I had not fully understood before: when people stand together, even great calamity can be met with courage and resolve. Now, nearly forty years later, as a new Tangshan rises where rubble once lay, whenever I think back to the earthquake, it is not the ruins I remember first, but the faces of those kind and brave people in the wreckage.

When the Disaster Struck

Zhiyu

In mid-July of 1976, I was reassigned from the Kaiping Sub Bureau to a task force formed by the Municipal Public Security Bureau's Public Order Division to handle a looting case. The work was urgent and demanding. I had not been home for ten straight days. On the afternoon of July 27, after lunch, Director He sent me back to Kaiping to report to the sub-bureau leadership and stop by my home. After one o'clock in the afternoon, I mounted my bicycle and set off from the municipal bureau toward Kaiping. It was the height of the dog days. The sun blazed overhead, heat rising in waves from the road. Along the way the crop leaves drooped, and even the grass by the roadside hung limp in the glare. When I arrived at the sub bureau and got off my bicycle, I noticed that the black rubber surface of the canvas raincoat clipped to the rear rack had softened in the heat. When I touched it, the melted rubber stuck to my fingers.

After work I went straight home and began cooking, washing clothes, and tending to the usual chores. My son was just over 50 days old. My wife was still on maternity leave. We had been married only a year and did not yet have a place of our own, so we were staying temporarily in my parents-in-laws' old tiled house, a place that had stood for more than a century. The heat that night was suffocating. It was hard to draw a full breath. To keep the mosquitoes away, my mother-in-law gave us a worn mosquito net to keep off the insects. Once inside it, sweat soaked through my clothes. It took a long time before I finally fell asleep.

At around half past three in the morning, the baby began to cry. My wife got up to change the nappy and feed him. At that moment, she heard a few dull thuds, *thump, thump*. She nudged me and said, "Listen! There's noise outside. Do you think someone's knocking at the gate?" I tilted my head and listened. All I could hear were chickens crowing and dogs barking. The noise was constant and chaotic. I wondered why the chickens and dogs were making such a racket in the dead of night. My wife finished feeding the baby, turned off the light, and lay back down.

Just then, the house began to shake violently. The beams, jars, doors, window frames, cupboards, mirrors hanging on the wall, and the bottles stacked on top of the cabinets crashed into one another in wild confusion. My wife screamed, "What's happening?!" I realised instantly that it was an earthquake and shouted, "Earthquake!" She cried, "Grab the baby!"

I leapt off the kang and turned back to pick him up. My hand had barely reached out when the thick earthen wall on the west side of the house collapsed, burying the three of us beneath it. Immediately afterwards, the beams and the rest of the house came crashing down as well. My legs were pinned below the edge of the kang, my body bent at a right angle across it. My wife was pressed face down, her head toward the kang. None of us had managed to reach the baby.

After the first violent shock passed, there was a brief moment when everyone was still stunned and had not yet come to their senses. All around us, it fell eerily quiet. I mustered every bit of strength I had, braced myself on my elbows, and tried to heave the weight off my body. I tried again and again, to no effect, it would not budge. I thought, if we couldn't get out, the baby had no chance of surviving. Despair overwhelmed me. I began shouting, "My child! My child! Help! Help!" My wife, however, was calmer than I was. She said, "Stop shouting. You'll run out of strength, and they won't be able to hear you outside."

Aftershocks continued to strike, and the weight pressing down on us grew heavier. Our breathing became shallow. At that critical moment, we suddenly heard my brother-in-law calling out, "Sister, where are you?" We answered at once, "We're here!" But outside it was too noisy. All around us, people were shouting as they searched for loved ones and cried for help. They couldn't hear us. "Sister, where are you?" he called again. We shouted back repeatedly, "We're here! We're here!" At last, they heard us. Locating our position, they began clearing the debris by hand, layer by layer. After about ten minutes, the weight above us was removed, and our heads and hands were freed. It turned out that my father-in-law's house had not completely collapsed. They had managed to crawl out and came straight to rescue us.

"Where's the baby?" they asked.

"We don't know."

"You go dig for the baby," they said. "We'll help the neighbours."

After they left, my wife and I first struggled to free our lower bodies from the rubble. Then we began, as carefully as we could, to dig toward where the baby had been sleeping. We knew he had lain between us. One by one we lifted away purlins, rafters, sun dried bricks, reed mats, clods of packed earth, broken tiles. Beneath all that, we found the worn mosquito net. In the darkness I pulled it aside. Suddenly I heard a choking cough.

Ah! He was there!

The sound cut through everything. We felt a rush of relief and gathered him into our arms. The worn mosquito net had collapsed over him, but it must have left a small pocket of space. That narrow space kept him from suffocating. We were also grateful that in the confusion no one had stepped on him. He would not stop crying. My wife tried to soothe him with her milk, but he could not nurse. When I reached out to touch his face, I found that his mouth, nose, and eyes were packed with mud. He was like a small clay figure.

What were we to do? The sky was thick with clouds. Not a trace of starlight. I could not see my own hand in front of me. There was no water to be found. My wife squeezed her milk over his eyes and into his mouth and nose, but that little bit of milk did nothing. It only mixed with the dirt and turned to mud on his face. We could not leave it like that. The mud must have been stinging his eyes, and his nose was so clogged he could barely breathe. We had to find another way. I lifted the child into my arms. I put my mouth to his eyes, his nose, his lips, licking and sucking the mud away bit by bit. It took a long while before he finally quieted down.

About half an hour later, the sky began to lighten slightly. In that dim gray light, I finally saw the state we were in. I had cuts on the top of my head, along the brow, at my jaw, and down both sides of my neck. Blood was running from them. My wife had suffered an internal injury to her waist and could hardly move. Our child had a split on his forehead that was still bleeding. I held the child in my arms and supported my wife as we climbed down from the ruins. Outside the doorway we found a patch of open ground and sat down. Rain began to fall. There was nowhere to hide, nothing to shield us from the wind or the water. I placed my wife and child in my mother in law's care, then stepped back out into the rain. I made my way to the Public Security Sub Bureau to report for duty, joining the earthquake relief work at once.

Time slips by, and youth fades before we notice. I have gone from being a young officer to a man past sixty. My son grew up, joined the army, and, after leaving the service, became a police officer himself.

Forty years have passed in what feels like a breath. Everything has changed. So much has receded into the distance. Only the Tangshan earthquake remains, etched deep in my life, a memory that endures.

Temporary Emergency Relief Centre

Yang Na

At 3:42 in the early hours of 28 July 1976, as Tangshan lay asleep, a deep rumble rolled through the darkness. Blue light flickered. The earth trembled. A devastating earthquake, one without precedent in a century and soon known across the world, struck. In a matter of seconds, buildings collapsed and entire streets fell flat. More than 240,000 living souls were gone.

At the moment the house collapsed, I heard a sharp, crashing roar, a shrill cascade of shattering. It was the sound of windowpanes breaking, mirrors splintering, cups smashing. It was the roof coming down. I was jolted awake and knew at once what had happened. The earthquake was upon us. Heavy beams and debris pinned me down. I struggled with all my strength, trying to rise, but I could not move at all. As the earth shook in violent waves, I felt myself buried deeper, pressed tighter. I could not open my eyes. I could not draw a full breath.

Suddenly, through the roar and dust, I heard someone calling out, "Dad! Mom!" The voice was faint and far away, yet I knew at once it was my younger brother, Yang Liu. He had practiced martial arts since childhood. He was a light sleeper, quick and nimble. At the first rumble from the earth, he had kicked open a window and leapt out to save himself. Hearing him call, I understood that I was not the only one trapped beneath the rubble. Our parents were buried as well. I was not afraid for myself. But the thought of our father and mother lying under the wreckage tore at me. They were the pillars of our family, the ones we leaned on. Grief surged through me. After that, I lost consciousness.

Guided by the two large locust trees in our courtyard, Yang Liu found the approximate spot where the house had stood. He shouted, and at last our father answered faintly from below. There were no tools at hand. He tried to lift a large slab of roof tile that was pressing down on our parents. It was enormous, nearly the size of the yard. He could not budge it with his hands, so he pressed his abdomen against it and pushed toward the courtyard, shifting his grip. As the slab shifted, he fell forward with it. A sharp edge tore into his abdomen, opening a deep gash, but he did not stop. After he moved the slab aside, there were still layers of mud, reed matting, rafters, bricks, and earth. He dug at them with his bare hands, clawing away one layer after another, until finally he pulled our parents free.

I no longer had the strength to call out. When Yang Liu came to pull me free, he moved with great care. He began at my head, clearing away the debris bit by bit. When my face was exposed, he scooped the mud from my mouth and nose with his hands. Air rushed in. My breathing eased, and my mind slowly cleared. Only after I was pulled out did I learn what had happened. Our father was not badly hurt. Our mother's waist had been broken. There was blood on our father's head. Our mother could not move at all. I felt light-headed, as if the ground were tilting beneath me. My limbs were weak and numb. Even if I had tried to move, I could not.

My brother saw that all around us the neighbours were digging frantically to rescue their own, and he did not stay to look after us. Ignoring the after-shocks, he ran to help others. To the east, Auntie Qi was crying out again and again for help beside a mound of rubble. Her husband was away in Beijing on a business trip. Only she and the children had been at home. She had managed to crawl out herself, but the children were still buried. She circled the ruins like someone trapped in a nightmare, helpless, not knowing where to begin.

Yang Liu rushed over, found the spot where they were pinned, and began digging with his bare hands. He tore a long gap through the debris. Along the edge of the kang, several heads slowly appeared. One by one he pulled them free, as if lifting turnips from the earth. Then he carried them, one after another, to a patch of open ground and laid them down.

There were four in all, three children and an elderly woman. When they began to breathe fresh air, they slowly regained consciousness. By some blessing, none of them was killed. By then, my brother's hands were already torn open, flesh and blood mingled, and blood ran from his fingers.

Across from us lived Aunt Yan's family. Her younger brother's voice rose again and again from the ruins, calling for help. Aunt Yan had been on the night shift. He had stayed behind to look after her five-year-old daughter. The adults were pulled free. By sheer luck, they were alive. But the little girl did not make it. By the time they reached her, she had already stopped breathing.

After I regained consciousness, I witnessed another scene that still tightens my chest. The aunt next door was kneeling over her eight-year-old daughter, giving her mouth to mouth. She kept at it for a long time. She refused to give up. She pressed and breathed into her child. The little girl lay quietly on the rubble, her face and clothes caked in dust. There was no response. She would never wake again. The mother's cries tore through the courtyard as she called out to her daughter, over and over. The girl's father, an engineering section chief at a machine tool factory, a man who doted on that child, was at that very moment helping neighbours nearby. In those minutes balanced between life and death, no one stood idle. People clawed their way free, then turned immediately to free their families and the people next door. When someone was found no longer breathing, there was no time even to settle the body. They had to let go and move on, racing toward anyone who might still have a chance. Looking back now, I understand how decisive those first hours were. The people of Tangshan did not wait passively. They saved themselves, and even more, they saved one another. That instinct to act without hesitation reduced the loss of life and bought precious time for the wounded who could still be treated.

What frightened us most were the constant aftershocks. At times there would be a sudden jolt, a heavy thud from beneath our feet, catching everyone off guard. At other times the ground would sway slowly from side to side, and the fear that had risen into our throats refused to settle. No one could relax. Rumors spread with the tremors. People said an even greater quake was coming, that the whole city would sink. Some were so terrified they ran to the railway tracks and clung to them, convinced that if the ground split open, the rails would steady them. Others said that the Douhe Reservoir might burst and the city would be swallowed by water... We had barely crawled out from death and were already facing fresh fear. The sky stayed overcast, rain falling on and off. There was no water supply, no electricity, no proper clothing, no food. The wounded groaned in pain. Bodies lay where they had been found. The aftershocks did not stop. The atmosphere of dread is hard to describe even now.

In that state of confusion, when no one knew what to do next, Uncle Zhao, Uncle Zhao, a Party member, gathered the other Party members who lived in the eastern section of our residential row and formed a temporary Party group. He then called an emergency meeting with one able bodied representative from each household. He reminded everyone that after the Xingtai earthquake, the central leadership had sent the People's Liberation Army to assist the rescue efforts. We had suffered greatly, he said, and surely help would come again. However, before the soldiers arrived, we had to rely on ourselves. We could not sit and wait. Under the organisation and encouragement of Uncle Zhao, people gradually steadied their emotions and regained their confidence. Everyone agreed that we would face the hardship together and not give in. Uncle Zhao made a rough division of responsibilities, covering the care of the injured, medical aid, daily living needs, and security.

Uncle Zhao was tall and thin, sharp minded and alert. He took charge of security. At night he led a few neighbors on patrol, keeping watch to prevent looting or other wrongdoing in the chaos. There were no shoes to wear. They walked barefoot over broken bricks and jagged debris, back and forth across the ruins. Their feet must have been cut and bruised, but they said nothing. Because of their efforts, the households nearby were protected, and people could rest a little easier in the dark.

Uncle Yang was responsible for daily living needs. This was by far the most difficult task. People had gone an entire day without water or food. There was a grain shop nearby, and someone suggested going there to take some grain. Uncle Yang refused. He said the grain belonged to the shop and could not be taken without permission. Later, someone dug up a few sweet potatoes. We lived on the edge of town, not far from a large sweet potato field. But people felt strongly that we should avoid harming the farmers' interests as much as possible.

Then several local children, Yang Liu, Qi Guoying, Ergou and a few others, discovered a large underground vegetable cellar just north of my home. It was a major storage site belonging to the vegetable company, and some people were already searching there for food. They carried back a basket of potatoes. In that moment, those potatoes meant survival. Later, water was fetched from the outskirts of the city. Using an aluminium rack from our backyard that had once held chicken feed, we set it over a wood fire and began boiling the potatoes.

The starving people were instantly energised. Hope returned to their faces. As the potatoes were nearly cooked, darkness fell. In the night, several children in their early teens approached us. When we asked them where they were from, we learned that all of them had lost both parents and had become orphans overnight. They lived nearby, in the Huaxin apartment blocks. As the children told the story of their families, everyone present was moved to tears. Kind-hearted Aunt Yang suggested that we must keep them with us and give them the first pot of cooked potatoes to ease their hunger.

From that moment on, we took in the children and shared whatever food and water we had. Later, the People's Liberation Army helped build simple shelters for them and took them back to their original neighbourhood. Only then did we say our tearful goodbyes.

After that, more and more people arrived or passed through. Our place gradually became a relief station. It was not until midnight that each of us finally received a potato that was only half cooked. My father was responsible for medical care. He had once served as a part-time doctor at the factory health station. At a time when the injured were everywhere and medicine was desperately scarce, he announced that there were medicinal herbs in our own courtyard. He had grown a large patch of traditional Chinese herbs himself, lush and green. They were called sanqi and were especially effective for treating external injuries. Remarkably, the herbs had not been buried by the quake. When people heard the news, they came in droves and quickly picked them clean.

My father also taught people how to tend to the injured, how to stop bleeding, and how to apply bandages. Despite his own injuries, he even dug through the rubble with his bare hands, found some medicine in our collapsed house, and used it to treat a child with a high fever, bringing the child back from danger.

The earthquake caused immense loss of life. The devastation was especially severe in the nearby water-machinery workshops where entire families were wiped out. From time to time, children came to my father, asking him to treat their parents' injuries. Their parents had been pulled from the rubble by neighbours, but they would never open their eyes again or speak to their children. In truth, the children had already become orphans. But they chose to believe that their parents were only asleep and would wake up. My father could not bring himself to tell them the cruel truth. They were still so young and still needed care. Yet reality could not be avoided. He could only explain things gently and ask neighbours on all sides to help look after the children. The children still had to go on living. There was still tomorrow.

Night fell again. A heavy stillness lay over everything. Uncle Zhao continued leading the young men on patrol. Uncle Yang was busy trying to find something for people to eat and drink. My father was still moving from place to place, tending to the injured. Several families gathered together, watching out for one another, offering what comfort they could. We waited and hoped for the People's Liberation Army to arrive. I leaned back against the trunk of the old locust tree and looked up at the stars. Sleep would not come.

Forty years later, when I recall that day, it feels at once impossibly distant and strangely close, like something seen through a dream yet still standing clearly before my eyes. 28 July 1976 — a date I will never forget.

The Day the Earthquake Came

Liu Ming

Forty years have passed, yet the scenes from that day in Tangshan remain as vivid as if it were yesterday.

When the earthquake hit, our house collapsed in an instant. My family of four was pinned beneath the scattered wreckage. A cement beam happened to brace against the large roof slab above our heads, leaving us just enough space to survive. My wife and I took turns calling for help. Above us we heard faint rustling. A small boy's voice drifted down through the rubble. "It's too big. I can't move it." With each aftershock, the space around us grew tighter. My breathing became short and strained. My wife said her leg hurt as if it had been broken. Our eldest daughter could only cry. From next door, where she too was trapped, Lu Jinghua called out from time to time to reassure us.

Through a crack in the debris, I could tell that daylight had come. Then I heard voices above. It was our neighbour, Guo Jianping. Soon others joined him. They all tried together, but they could not shift the heavy slab. Someone found a round wooden pole. They began prying little by little, wedging it upward bit by bit. Yet the higher they lifted, the more the slab pressed down toward my side. They must have heard my breathing grow harsher, because they stopped. After a brief discussion, they called down to ask whether they should try pulling us out by force. Struggling for air, I answered, "Do whatever you have to. As long as we're all alive." One by one, they pulled my daughter, my wife, and then me out into the light.

My younger daughter was not beside us. At first, we feared the worst. In the confusion we thought she might already be gone. Then, from somewhere in the rubble, we heard her murmuring in her sleep. The sound set everyone moving again at once. They dug frantically, searching for her voice. She was the last to be pulled out, and only when they lifted her free did she finally wake. It turned out that when the house collapsed, the mosquito net had wrapped around her and left her hanging just below the edge of the kang. She had not been hurt at all.

Five people helped pull us out. Four of them were complete strangers. Before we could even catch our breath or ask their names, they were gone. Only later, after asking around, did we learn that they were residents from nearby. Two of them had once studied at No. 19 Middle School.

We longed anxiously for news of the rest of our family, yet our injuries kept us from moving about. My third younger brother arrived only after we had been pulled out. He had always been the one neighbours turned to when trouble arose.

He had just crawled from his own collapsed house when a girl from farther down the lane, Hou Yaya, ran to fetch him. He went straight with her and did not stop until he had dug her parents out from the rubble. Only then did he return to help our family bury our second younger sister. After that, he hurried over to us without a pause. He dropped off two sheets of plastic, saw that both my wife and I were injured, and took our eldest daughter away to safety.

When the earthquake struck, my father-in-law and brother-in-law broke out through the east room window. My mother-in-law and my wife's younger sister had been sleeping in the side room. When that room collapsed, the roof gave way and the night sky opened above them, stars scattered across the darkness. Miraculously, neither of them was hurt.

All around us, cries for help and the sound of weeping rose from every house. Faced with a disaster none of us had ever imagined, my brother-in-law made a clear and resolute choice. Without hesitation, he threw himself into rescuing others. One after another, he helped pull seven people from the rubble, some of them joining in to rescue others once they had been freed. He went a long way from home to find water. On his way back, those he passed were so desperate that the water was shared and quickly gone. He turned back without complaint and went searching again. By the time he reached us, it was nearly dusk. Seeing that we were not in immediate danger, he did not stop to rest. He set out once more to check on my second and third younger sisters, who had been sent down to Quzhuang in Kaiping.

The aftershocks did not stop. No one dared return to their own ruined homes to search for stored grain. Guo Jianping managed to bring back a few scallions and some small green tomatoes. We shared them and quieted our hunger as best we could. Adults could endure an empty stomach. An infant not yet a month old could not. The baby's mother, Qiang Huzhen, had died in the quake. The child cried without pause from hunger. The young father stood there drenched in sweat, helpless. My wife's waist and legs were injured and she could hardly move, but we had someone bring the baby to her. Though she was in pain, she let the child nurse. Her nipples ached from the sucking, and our own little daughter began to cry as well. But in that moment, the crying of our daughter helped quiet the orphaned infant's wailing. The two sounds seemed to hold each other.

Together with five neighbouring households, those who were still able moved about gathering what we could. Piece by piece, we found scraps and boards and cloth and set up a lean-to by the roadside. Behind where we sheltered, on the broken floor of a collapsed two-story building, a victim's leg hung down from the edge. Not far to the east, where the road turned north toward the Baima Mountain cemetery, the dead were carried past us one after another, taken toward that distant ground. As survivors, we could do little but bow our heads and pray. To those who had shared this earthly life with us and were now leaving it, we wished them peace on their final journey.

Night fell, and heavy rain poured down in sheets. The shelter we had thrown together in haste could block the sun well enough in the daytime, but against the rain it was nearly useless. Water streamed through the seams of the plastic sheets overhead. We rushed about in the downpour, grabbing whatever could help. Some stuffed rags and bits of debris into the gaps from outside. Others stayed inside, holding out basins and buckets to catch the dripping water. Luckily, when we laid down the wooden boards for the floor, we had propped them up slightly with bricks. The rainwater ran beneath instead of pooling where we lay. Gradually the rain eased. Sister Song Baozhu brought out all the quilts her family had and spread them without hesitation. Some were laid beneath us, others used as covers. People shifted and made room for one another. The center was given to the children and the women; the edges were left to the men. After a day of terror and relentless strain, more than a dozen of us lay crosswise and lengthwise in that narrow space, and at last, exhaustion carried us into sleep.

Time passes like a dream. What once felt immediate now drifts like smoke, yet that day remains sharply etched in my memory. The people who pulled my family from the rubble, the neighbors who shared that fragile shelter with us, have long since scattered in different directions, each living their own life. There is only one I have never met—the little boy who first began digging above us on the mound of debris. I never saw his face. By now he must be a grown man, I imagine. I like to think he has carried that same kindness into adulthood. I hope that the compassion born in the midst of such catastrophe endures, and that the quiet strength of human feeling continues to take root wherever it can.

An Unforgettable Recollection

Zhongjian

In the early hours of 28 July 1976, I was sound asleep when the ground began to shake. Even before I fully woke, the walls were collapsing in my dreams. By the time I opened my eyes, the heavy roof of our single-story house had already fallen. We were fortunate. My wife and daughter were not trapped beneath the beams. They suffered only minor cuts to their heads and legs. My son and I, however, were pinned under the roof. A wooden chest that had been placed at the head of the kang bore part of the weight. That chest kept enough space for us to survive.

At the time, I was living in the Power Bureau residential compound at Majiatun, on the east side of Xiaoshan, less than a hundred meters away. From there toward Fengnan lay the heart of the earthquake zone, where the destruction was especially severe. In our entire compound, only one row of houses remained standing. Everything else was reduced to rubble. It did not take long for those who were still alive to begin rescuing one another. My wife and daughter were dug out by neighbours, who immediately rushed off to help others. They broke a hole through part of the collapsed roof, and I saw daylight filtering down. I told my son to climb out first. When I tried to shift my body, I realised my left arm had been fractured. There was no open wound. The roof must have struck it bluntly. The opening was near my legs, and those above could not pull me through. I had to twist myself upward using only my right arm and my head for leverage. Bit by bit, I managed to raise myself until my head cleared the debris. I told them not to drag me up. I needed to protect the broken arm so the bone would not pierce a blood vessel. I asked the children to find two small wooden boards and tie them together with rope. That served as a temporary splint.

At around six thirty in the morning, the time for the Central People's Broadcasting Station's early news, we suddenly heard music drifting from the direction of Tangshan No. 2 Middle School. Everyone was startled, then excited. How could there still be electricity there? Who had turned on the loudspeaker? The familiar broadcast voice rang clear across the ruins. It brought hope. Beijing had not been struck. The central leadership would send help. As those words sank in, many of us found our eyes clouding with tears.

We had gone a full day without eating, and the children were hungry. Somehow, they found empty plastic bottles and brought back water from beneath the water tower at a nearby paper mill. Someone managed to dig out a sack of white flour from their collapsed home. Another found some dried vermicelli. A cooking pot was salvaged. Firewood was gathered. In the middle of the courtyard, everyone worked together. They boiled the vermicelli and cooked a rough pot of dough drop soup. It became our first meal after the earthquake. Half the families had lost someone. There was no time, no strength, and no clear way to attend to the dead. People could only wipe their tears in silence. One girl whose surname was Gu cried loudly for her little brother. She refused to eat, and no one could persuade her. There were also Engineer Zhang, his wife, and their daughter, all in the prime of life. They were buried beneath the rubble. People called their names again and again. No sound came back.

On the second day, those who had ventured farther out returned with alarming news. They said the embankment of the Douhe Reservoir had cracked and that we had to evacuate at once. People scattered in different directions, each thinking only of escape. Later we learned that the dam had not broken at all. It had been nothing more than a false alarm.

On 30 July, someone suggested that we move to the Yutian Power Supply Station. Perhaps the destruction there was lighter. After some hurried discussion, our whole family decided to head for Yutian. Before leaving, I endured the pain in my injured left arm and turned back for one long look at what had been our world. To the south of Xiaoshan, only a flour mill remained standing. All the other buildings had fallen into ruins.

We moved slowly along the road. When we reached the east gate of the Qixin Cement Factory, a medical team from the People's Liberation Army had already arrived. At the sight of them, something in me gave way, and tears rose before I could stop them. They replaced the rough splint on my injured arm with a proper one. For the first time since the earthquake, I felt that someone had truly come for us. At the Gangyao Power Supply Station, we managed to find a small diesel vehicle. Seven of us squeezed onto it. Just before climbing aboard, someone found a sack of flatbread that had been dropped from an aircraft. We broke it apart, each taking a piece. We ate as we started down the road. The thirst along the way became unbearable. We picked up discarded liquor bottles and enamel mugs and scooped water from roadside puddles. We knew it was not clean, but there was no other choice.

When we reached the Yutian Power Supply Station and saw that the buildings and substation there stood intact, we could hardly believe our eyes. A surge of relief filled us. After so much ruin, the sight felt almost unreal. The comrades at the station wasted no time. They prepared food for us, and we sat down to a real meal for the first time in days. Soon afterward, they arranged to send us to nearby Yao Bazhuang Middle School. Tents had already been set up there. Straw had been spread on the ground for bedding. There were cooks preparing meals and doctors sent from the county hospital.

At last we had a place where we could lie down and rest without fear, a temporary home made up of fifty or sixty people bound together by hardship. Food was no longer a pressing worry. We had steamed buns and vegetables three times a day. There was medical treatment. Some of the injured had already been settled there. The next day, my left arm was set in plaster. The director of the repair division, Li Guochang, called out to me. I almost did not recognise him. His head was swollen to the size of a basketball. He told me that in his family only he and his daughter had survived. A sharp ache went through me as I listened. Later, I heard that he did not survive the transfer and passed away on the way to Wuhan.

About the third day, several sedans drove into the yard of Yao Bazhuang Middle School. The first person to step out wore a white headscarf wrapped around his head, dressed in plain farmer's clothes. His face was weathered, dark with a flush beneath the skin, the lines on his forehead deep and clear. I recognised him at once. It was Chen Yonggui, then Vice Premier of the State Council and a national model worker in agriculture. He walked over to us and, in a heavy Shanxi accent, said, "Is everything ok?" No one answered. At first I did not quite catch his words. When he repeated, "Is everything ok?" I understood. My hometown was in Henan, and I had lived in many places, so I replied, "It's all right." He seemed pleased that someone had understood him and smiled at me. He then gestured toward a woman standing beside him, of above average height, wearing a simple cloth sun hat shaped almost like a straw hat. She looked like a modest young university graduate. "This is Comrade Wang Hairong," he said. "We are here on behalf of Chairman Mao, the Party Central Committee, and the State Council to visit you. We hope you will take good care of your injuries, overcome this natural disaster, and rebuild Tangshan." Many people were moved to tears. Some could not even speak. Wang Hairong took out a handkerchief and helped wipe the tears from my face and from those next to me. Chen Yonggui and the others shook hands one by one with those in the front row before leaving. They still had other resettlement sites to visit.

After the earthquake, makeshift shelters sprang up everywhere, built against aftershocks and against the wind and rain. Inside those shelters there were electric lights again, the smoky smell of cooking fires, the clatter of pots and bowls, the quiet closeness of families who had survived together. People had a home once more. Our work unit put up a row of shelters on an open stretch of land southeast of Nanliutun. I moved there to live. The materials all came from aid sent across the country. The soldiers and the people of Tangshan who had survived became builders without instruction. The process was simple. First, pits were dug and cut wooden poles set upright. Cross-beams were laid across them. A layer of reed matting was spread over the roof, then covered with tar felt and weighted down with salvaged bricks or stones. That was enough to give the structure form. Most of the roofs were sloped. The lower part of the walls was built from bricks retrieved from the ruins, stacked to about a meter high. Above that, reed matting was used again. To keep out the wind, the mats were plastered with mud. There were no courtyards. The beds were nothing more than wooden planks nailed together.

To keep warm through the winter, I would take my coal ration booklet, get on my bicycle, and roam around the city searching for lumps of coal to burn in the stove. One time, at the coal yard near the Northwest Well, I finally found what I had been looking for. I borrowed a flatbed cart, bought a full load, and pulled it all the way across the entire city of Tangshan before I could bring it home. Some of the coal pieces were so large that I had to chop them into smaller chunks with an axe before they would burn properly. When they caught fire, the iron stove lid would glow a deep red from the heat. To keep the warmth from escaping, we extended the stovepipe, adding more sections to make it longer. Even in the depths of winter, the room would grow so warm that we found ourselves sweating. In that carefully preserved warmth, there was a quiet delight, a simple yet deeply felt joy that rose naturally from the texture of everyday life.

Inside our fenced courtyard, I also grew sunflowers and kept chickens. The yard was too small for vegetables, so I planted a dozen or so sunflower stalks along the fence. The water came from a public pump. I carried it back in buckets and measured it out carefully, one generous ladle for each plant, without favouring any of them. They had to be watered once a week without fail. For fertilizer, I used chicken manure.

At the time I kept three chickens. I would chop up vegetable leaves, mix them with cornmeal, and feed it to them. Occasionally, I tossed in a handful of sorghum grains as well. One day, the chickens flew out of the yard and wandered over to a garbage heap to scratch for scraps. They had no way of knowing that the garbage heap had been sprayed with dichlorvos to disinfect it. One after another, they kicked a few times and went still. I could only watch, helpless. After that, I did not keep chickens anymore.

The sunflowers, however, thrived. Their stalks grew thick and sturdy, and the seed heads were as large as washbasins. The public water pump was more than a place to fetch water. It was where people gathered. Women stood there with large basins, washing clothes and exchanging bits of neighbourhood news. Men who lived farther away carried water home with shoulder poles, while those who lived closer lifted two buckets by hand. With one bucket in each hand, they would break into a quick trot. If they moved too slowly, the buckets would knock against their legs and the water would spill over the rim. Meanwhile, the Ministry of Electric Power instructed the North China Electric Power Administration to establish an earthquake relief command center without delay. Working around the clock and undeterred by hardship, they raced against time. They led electricians from across Northeast China and Hebei Province, together with the surviving staff of the local power supply bureau. Only fifteen hours after the earthquake, electricity was restored to the provincial and municipal earthquake relief headquarters. Soon afterward, the Longwangmiao Waterworks regained power as well. As the damaged pipelines were gradually repaired, even the makeshift shelters throughout the area once again had running water.

I Was in Tangshan When the Earthquake Struck

Kou Zhen

In April of the year the earthquake struck, I was transferred to the Tangshan Air Force Station, where I lived in the main administrative residential compound. On 27 July, the weather was stifling. Not a single leaf stirred. The air felt thick and motionless, as if it had solidified. That night, I tossed and turned, unable to fall asleep. I do not know how much time passed before I drifted into a hazy doze. Suddenly, my bed began to shake violently. I opened my eyes and saw flashes of light flickering outside and I heard rolling thunder. My first thought was that enemy missiles were incoming.

At the time, international tensions were high, and we lived in a state of constant readiness, the string of war drawn tight. I immediately slid under the bed for cover. Then, in a split second of lightning, I dashed out of the room. No sooner had I reached the doorway than a deafening crash split the air. The building I had been living in collapsed behind me. It was a narrow escape. Had I stepped out even a few seconds later, I would have been gone from this world.

To the City for Rescue

At the emergency meeting held at the air station, I was assigned to lead more than twenty soldiers into the city to assist with rescue operations. By around 5:10 a.m., everyone had assembled, and we set off immediately by truck. When we reached the bend near the Northwest Well, the vehicle was stopped. People rushed forward, lifting the injured onto the truck and pleading for it to turn back and take them to the airport medical unit. We had no choice but to jump down and continue on foot.

All along the way, the city lay in utter devastation. Buildings had collapsed into heaps of rubble. Cries and groans drifted through the air. The dead and wounded were everywhere. The sight was almost too harrowing to take in. We began our rescue work in front of a row of collapsed buildings on the west side of Wenhua Road. From the scattered army-green quilts and bedding among the debris, we realized this had been the guesthouse of the Air Force's Sixth Corps. Casualties were severe. Some of the injured were moaning weakly; others were trapped between slabs of concrete and stone, already lifeless. I divided the men into two groups. One group worked to pull out the wounded. The other searched for those still buried underneath. When we heard faint cries for help from beneath the rubble, I shouted at the top of my lungs, "Someone's down there. Start digging!"

We quickly cleared a small opening. Shining a flashlight inside, we saw that the trapped man was more than ten meters away, beyond our reach. A thick concrete slab blocked the way. The air inside was stifling, and he was pinned so tightly that he could not move. At any moment, he could suffocate. We were frantic. Aftershocks kept shaking the ruins, and the hole we had just dug would collapse again, clogged with bricks and chunks of masonry. We cleared it, only to have it blocked once more. Just then, I noticed a car parked not far away. If there was a car, there had to be a jack. Without stopping to ask whose it was, I ran over and took it out. A thin, wiry soldier snatched it from my hands and said, "The opening's too small. You won't fit. Let me go."

What a brave young soldier he was. We stayed outside the opening, fanning air into the hole and shining our flashlights to guide him. Inside, he wedged the jack beneath the concrete slab that was pinning the man down and began cranking the handle. Slowly, almost unbelievably, the slab lifted a few centimetres. We passed a backpack strap through the gap and told him to tie it securely around the man's ankles. Once he backed out, we began to pull, inch by inch, careful and steady. Eventually, we drew the man free. We told him to keep his eyes closed. He lay there gulping for air, breathing in great, ragged gasps, unable to speak a single word. Later, we learned that he was the guest-house manager.

Burying the Dead

On the night of 28 July, it rained again. The rain was light, yet it felt as though salt were being sprinkled onto the wounds of the injured. By the 29th, the dead lay scattered across the airport grounds. Flies swarmed everywhere, and a foul odor hung heavily in the air. To prevent the outbreak of epidemic disease, the command headquarters ordered immediate burial. Once again, I was put in charge. We divided the work carefully. Some were assigned to dig graves, others to transport the bodies. I was responsible for registration. I recorded in detail where each person had been found, what clothing they wore, their gender, approximate age, and the exact burial location, so that relatives could later make inquiries.

The task appeared straightforward, but it was anything but easy. Many of the soldiers had never seen a dead body before, let alone those who had died from injury. I myself felt a measure of fear. Yet when we reminded ourselves that the deceased were ordinary people, and that we were soldiers of the people, duty-bound to treat them with compassion and respect, our hesitation fell away. Through constant encouragement and mutual support, we overcame both our fear and the overwhelming stench. In a single day, we buried 108 bodies. We were so exhausted that we could hardly eat.

I kept the small notebook in which I recorded the details for several years afterward. It proved more useful than I ever expected. When families came searching for missing loved ones, I would consult my notes and describe what

I had written down. If the information matched, I felt that the little notebook had truly served its purpose. The hardship and fatigue seemed worthwhile, knowing that the families of the dead could at least find some measure of closure. I remember a cadre from Yunnan who had come to Tangshan for a meeting. He was injured in the earthquake and later died on the square in front of the main auditorium at the airport after being transported there. When comrades from Yunnan came to look for him, I received them personally and explained how his body had been wrapped and handled. They listened quietly, and when they left, they seemed reassured.

Indelible Memories

Baohua

The Tangshan earthquake that shocked the world took place more than forty years ago, yet the scenes from that time remain etched in my memory. They return with startling clarity, as if everything had unfolded only moments ago.

In July 1976, I was transferred from Erbozi Commune in Qianxi County to Tangshan and appointed Deputy Director of the Tangshan Regional Bureau of Agriculture. I was supposed to assume my post in mid-July. However, on 18 July, I attended a training program for young cadres organised by the prefectural Party committee. The program was held in Qishuzhuang Commune, Fengrun County, about 25 kilometres from Tangshan, and was scheduled to last twenty days. I was instructed to take up my appointment only after completing the training.

During that period, three colleagues and I stayed at the home of Chen Xisheng, the Party secretary of Chenjiapu Village. About a week into the program, we began noticing unusual signs. The well water turned cloudy. By the evening of the 27th, the changes were even more striking. Rats darted about in agitation. Chickens flapped wildly, dogs barked and leapt around, and cattle and sheep lowed in restless distress. None of us understood that these were warning signs of an impending earthquake. On that sweltering night, unaware and unsuspecting, we went to sleep.

When the earthquake struck, a deafening crash tore through the darkness, followed by a sudden flash of light. The ground heaved violently beneath us. The gable wall of the house where we were staying collapsed. Fortunately, rural homes were built with interior support pillars. The wall fell outward, and the roof did not cave in. Of the four men sharing the room, one was hurled across the space and ended up pinned beneath a wooden cabinet on the far side of the courtyard. Two others and I scrambled through the window into the yard. In the confusion, I grabbed a garment from the clothesline and hastily pulled it over myself, at least to cover my body. I felt embarrassed to find myself dressed in a woman's clothes. I then slipped back inside and changed into my own.

In the early hours of 28 July, just as dawn was breaking, more than thirty of us from the training program set out for the city. We were led by Lu Xiaoxia, Deputy Director of the Prefectural Organisation Department. In a light rain, we climbed onto a village wheeled tractor and hurried toward Tangshan. What met our eyes was beyond anything we could have imagined. The entire city had been levelled. Everywhere lay heaps of rubble. Great fissures split the roads open. Bodies were scattered in every direction, and streaks of bright red blood stained the ground. From time to time, faint cries for help drifted up from beneath the ruins, mingled with the piercing sobs of children.

A crushing weight settled in my chest. The grief was almost unbearable. Aftershocks struck again and again, shaking the ground so violently that none of us could stand steady. There was no water, no food, no electricity. Hunger and thirst gnawed at us. And beneath it all was a fear that defied words, as if the very sky might collapse at any moment.

Our task was to rescue those trapped in the ruins of the regional commercial building. Many out-of-town guests had been staying there. As long as there was even the slightest chance that someone was still alive, we were to do everything we could to save them. Among the thirty or so of us, about one-third were women. The men climbed onto the rubble to carry out the rescue work, while the women remained below, directing us and keeping watch. Whenever aftershocks struck, they shouted for us to come down immediately to ensure our safety.

The tragedy was that we had no tools. We worked with our bare hands. Faced with reinforced concrete and heavy prefabricated slabs, we were powerless. All we could do was move loose stones and bricks. At times, we accidentally stepped on the head of a corpse. There was no time to dwell on it. Saving the living came first! Our minds were fixed on one thought: rescue. Clothes were torn, feet pierced by debris, fingers scraped raw and bleeding, yet none of us felt the pain. Just before the five o'clock aftershock that afternoon, we finally pulled out a young waitress. From our brief exchange, we learned her name was Xiao Jin, from Laoting County. She had suffered only minor injuries. When we carefully brought her out, we realised she was naked, her clothes having been torn away in the collapse. She was in her early twenties, tall, with a square face and long eyelashes. Several of us quickly wrapped her in a bedsheet. A few of us grown men quickly wrapped her in a bedsheet, then slowly carried her down from the rubble and laid her aside. Only then did we allow ourselves a faint, hard-earned smile.

There was a man buried deep beneath the rubble who shared my name, Zhang Baohua. He was the political commissar of the Chengde Military Sub-district. When we called out to him, he answered. Our voices reached each other clearly. We could determine his approximate location, yet we could not see even the slightest trace of him. His orderly, a capable young soldier, had spent the night at a relative's home and escaped injury. The young man pleaded with us again and again, begging us to rescue his commissar. All we could do was speak to the man beneath the ruins, offering reassurance and urging him to hold on, promising that we would bring him out. He answered that he would wait patiently for us. But at five o'clock that afternoon, after another violent aftershock, no matter how loudly we called, there was no reply. The brutal tremor had taken his life. Only moments before, he had been speaking to us. Then, in an instant, he was separated from us by life and death. In that moment, I felt a shock deep within me. How fragile a human life is. A person can be alive and conscious one minute, and gone the next.

To this day, I still carry a sense of guilt that we were unable to save that commissar.

Although we had gone the entire day without food or water, we endured our hunger and thirst and continued the rescue work. Night fell. The rain stopped, but aftershocks continued to assert their force. In that moonless, lightless darkness, we stood amid the ruins at a loss for what to do next.

At some point, an order reached us from a superior, instructing all of us to return to our original units and await further tasks. Communications were completely paralysed. Telephones were useless. Only the routes toward Beijing, Zunhua, and Qianxi remained open; roads leading toward Tianjin and Qinhuangdao were cut off. It was not until after ten that night that I finally made my way back to Qianxi County. Having eaten and drunk nothing all day, I was faint with hunger and thirst. I swallowed a few bites of food and then shared a tent with Xiao Peihui, the director of the Qianxi County Hospital.

As I described the day's rescue efforts to him, he suddenly cried out in astonishment. The young waitress we had pulled from the rubble, Xiao Jin, was his younger sister. While he had been anxiously waiting for news of his sister, I had unknowingly brought him word as if sent by an angel from the heaven.

He exclaimed with joy, "Ah! How could such a coincidence happen in this world! What an extraordinary twist of fate! Thank you, truly!"

"You're giving us far too much credit!" I spoke. "This was our duty! We had no idea Xiao Jin was your sister. It just so happened!"

The next day, 29 July, I returned to my original post at Erbozi Commune to hand over my responsibilities. Before the newly appointed Party secretary had even arrived, I received notice to report on 1 August to the Prefectural Organisation Department and then proceed immediately to take up my post at the Regional Bureau of Agriculture. The newly appointed bureau chief, Sun Chang, welcomed me warmly. He arranged for me to stay in a tent set up outside the seed warehouse. It was the peak of summer, the hottest stretch of the year. The tent was damp and stifling; the moment one stepped inside, sweat soaked through one's clothes. Fortunately, airplanes flew overhead each day spraying disinfectant, so at least we were spared the torment of mosquitoes and flies.

Among the many things Director Sun told me, one in particular has never left my mind. On the night of the earthquake, twenty-two people had been inside the Bureau of Agriculture dormitory building. When it was flattened, seventeen died instantly, including the former bureau chief Zheng Xin and deputy chief Yang Chunqi. Of the five who survived, only one suffered minor injuries. The other four, including two deputy chiefs, were severely injured and had already been transferred elsewhere for treatment.

What he told me about his own family was almost beyond belief. They lived on the fourth floor. Of his three daughters and one son, the eldest daughter died in the earthquake. The others sustained only minor injuries. The youngest daughter, just in her early teens, seemed as though heaven had spared her deliberately. The earthquake had flung her, bed and all, from the fourth floor down to the ground. She was still lying on the bed when she was found, completely unharmed. After hearing his account, I was struck by the incomprehensible force of that earthquake. In the days following the disaster, with roads blocked and rescue teams unable to reach the area, people were so desperate for water that they drank even the bathwater from public bathhouses until it was gone.

From this, I could not help but reflect to myself: the young cadres' training program had, by sheer chance, spared me from that catastrophe. I had survived a great disaster. Beyond that, I could only say it was an extraordinary coincidence. The years that followed felt, in a sense, like a gift from heaven. Since the earthquake had not taken my life, I believed there must still be things left for me to accomplish. And quite naturally, I found myself viewing the gains and losses of this world with a greater sense of detachment.

After I took up my post as deputy director, I worked more like a production team leader than a government official. During the day, I led everyone into the ruins, digging for survivors and salvaging what we could, then putting up makeshift shelters. It was even more exhausting and more taxing than farm work. We survived on biscuits, dried shrimp, and other provisions airdropped by the state. The water we drank was delivered by military trucks.

Work was relentless and exhausting. With every sector paralysed and urgent matters pressing in from all sides, there was no time even to write home. My elderly parents and my wife had no idea whether I was dead or alive. They could only worry. About a month after the earthquake, I finally managed to seize a brief moment to return home. When my family saw that I was safe and uninjured, they were visibly relieved. My mother kept repeating the same words over and over: "Thank heaven! Thank heaven!"

That October, a construction company was recruiting workers, and my wife signed on as a construction labourer. She moved in with me into a makeshift shelter, its brick walls topped with tar paper roofing. It was crude but sufficient. At the Prefectural Committee's communal canteen, we would see the Party secretary, the commissioner, and the heads of various committees, offices, and bureaus standing in line alongside everyone else. They ate the same food we did, cornmeal porridge, steamed corn buns, scallions mixed with dried shrimp. In daily life, everyone was treated the same. In the face of such a disaster, no one enjoyed special privileges.

After we moved into a semi-permanent shelter and began to settle into a life of our own, we naturally found ourselves short of many everyday necessities. Wang Changwu brought over a kitchen knife. Cui Xueli carried in firewood for us. Wang Guijiu delivered a small chopping board. Whenever I was away on business and unable to return home, Lin Gao and his wife would send their teenage daughter, Ruzi, to keep my wife company. Each time she came, she brought two large apples, making sure one was for my wife. My wife had grown up in the countryside, raised on sweet potatoes, and in earlier years apples had been a rare treat. In those days, to bite into a sweet, crisp apple felt like a small but genuine pleasure. My wife worked long days at the construction site, coming home dusty and exhausted, and still had to cook for herself. When loneliness crept in, our neighbour, Sister Guo Shumin, would drop by to chat and keep her company. Over time, the two became close confidantes.

All those small acts of warmth, added together, became something vast and deeply human. After enduring the trial of the earthquake, the people of Tangshan seemed gentler, purer, more open-hearted. Surrounded by the care and warmth of those around us, we felt an unmistakable happiness rising from within.

Living the Earthquake, Witnessing Rebirth

Shuwen

The Tangshan earthquake left a deep mark on me. The Tangshan earthquake left a deep impression on me. Even now, when I think about it, I still feel a lingering sense of fear. To be honest, I would rather not revisit those painful days. The memories are difficult. Yet I feel I should share them, so that others, especially those who came after us, may understand what we went through. Life as it is today was not easily won. Perhaps remembering those years can remind us to treasure what we have now, to treat our families and friends with kindness, and to live with gratitude in our hearts.

Forty years ago, I was just an eleven-year-old girl. I had never even heard of an earthquake. In the early hours of that morning, I was in deep asleep when a violent jolt shook me awake. The bed lurched beneath me. I was stunned, unable to understand what was happening. Instinctively, I grabbed the bed rail with both hands. In the darkness, I stared in confusion and fear at the house swaying and trembling around me. Then I heard a sharp crack as the clock on the table fell and shattered. A split second later, something seemed to come crashing down over me, overwhelming and unstoppable. After that, I remember nothing.

When I came to, it was already broad daylight. Something heavy was pressing down on me, and through the haze I could hear cries and shouting outside. Summoning all the strength I had, I screamed, "Help! Somebody help me!" Before long, a young man from next door managed to dig me out and carry me on his back to the roadside. By chance, a passing vehicle stopped and took me to the square outside my father's workplace. Many people had already gathered there. A light rain was falling. I was placed beneath a parked cargo truck to keep me out of the rain.

I do not know how much time had passed before I saw my aunt's husband and my youngest uncle standing in front of me. They both lived in the countryside of Fengrun, and their own homes had suffered damage, yet they had left their families behind and hurried over to find us. My uncle said quietly, "Your sister has been pulled out too. She lost too much blood. She didn't make it." His voice was calm, but to me the words struck like a bolt of lightning. My sister had been talking to me while we were buried in the rubble. How could she suddenly be gone? I refused to believe it. I had to see her. My uncle tried to stop me. "Don't go," he said. "There are too many who didn't survive. They're everywhere."

Just then, a neighbour from downstairs brought my nine-year-old brother over to me. While searching for his own parents, he had discovered my brother buried in the debris after falling from the upper floor and managed to rescue him. The neighbour's parents, however, had both perished in the earthquake.

My brother was covered in cuts and wounds, still bleeding from several places. As I looked at him and at everything around us, tears burst from my eyes. Toward evening, my uncle said quietly, "We still haven't found your parents. Let's take you and your brother to your aunt's home first." The roads were split open by deep cracks. A bicycle could not be ridden; it had to be pushed along. I was placed in a large wicker basket. Each jolt sent a sharp, piercing pain through my leg. Around midnight, we finally arrived.

My aunt's house had not collapsed. In the courtyard, they had put up a simple A-frame shelter. At night, the whole family crowded into that small space. The air inside was stifling, and mosquitoes tormented us without mercy. My aunt stayed awake, fanning us to keep the insects away. My older cousin helped take care of my toileting, and every day she carried my brother to the clinic to have his wounds dressed. I remember all of it clearly.

Four days later, my leg had swollen as thick as a post. My eldest uncle said that their commune was sending several severely injured people to Beijing for treatment, and they would arrange for me to go as well. There were no proper vehicles available. The commune dispatched a large tractor, and the clinic director, together with medical staff, accompanied us. My uncle left his own family behind and came along to see me through.

We set out at night. Along the way, we encountered countless stretches of cracked road that were impossible to pass, forcing us again and again to take detours. After what felt like endless jolting and bumping, we finally reached Beijing, where we were admitted to Beijing Chaoyang Hospital.

The medical staff hurried to examine me. The diagnosis was a fracture of my left femur. I needed traction immediately. There were so many fracture patients that the hospital ran out of the metal weights normally used for traction. Bricks were used in their place. Even with anaesthesia, each hammering motion sent a piercing pain through me. I clenched my teeth and endured it. No matter how badly it hurt, I refused to cry. The doctors and nurses praised me for being brave.

On the fifth day after the traction was set, Beijing began to experience aftershocks as well. To prevent those of us who had just survived one disaster from facing danger again, the hospital decided to transfer us to a safer place for continued treatment. This time, we were transported by train. I was carried on a stretcher and lifted through a train window into the carriage. At every stop along the journey, crowds gathered to bring us letters and gifts of sympathy. At every station, banners stretched across the platforms bearing words such as "Respect to the people of Tangshan!" It was on that journey that I first understood, in my own childish way, how people come together in times of hardship.

After traveling for what felt like an endless stretch of time, the train finally came to a stop. Through the window, I saw a line of People's Liberation Army soldiers standing in formation on the platform. A light rain was falling. We were lifted out of the carriage and placed onto an open military truck waiting there. Four soldiers stood at the corners, each holding up a white sheet above us to shield us from the rain. We lay beneath it, dry, while they themselves were soaked through. We were taken to Hospital No. 417 in Lintong County, Shaanxi Province. The surroundings were beautiful, with pomegranate trees lining both sides of the road. A few days after we were admitted, word spread that aftershocks had been felt there as well, and the hospital staff moved us under our beds as a precaution. The jolting and shifting during the transfer had caused the bone in my leg, which had not yet fully healed, to slip out of alignment again. The doctors said I would need traction once more. I had to go through the painful traction procedure once again.

Around November, the hospital informed us that we could begin preparing to return home. We were overjoyed. Before our departure, the hospital held a small farewell gathering for us and provided each of us with brand-new cotton-padded jackets and trousers, along with some local specialties. When we first arrived, we had been carried in on stretchers. Now, each of us was able to walk out of the hospital on our own. Through the train window, I looked at the doctors and nurses who had cared for us with such devotion over those months. A deep reluctance to part rose in my heart. It was they who had helped me stand again. It was they who had given me the courage to keep living. As the train slowly pulled away, I waved to them, my eyes already blurred with tears.

When the train pulled into Tangshan Station, a child who had been separated from her family for months was finally coming home. I longed to see the place I remembered, to see the faces I had missed so dearly. But when I stepped out of the station, I froze. The streets were lined with low, irregular rows of makeshift shelters. Bricks weighed down sheets of tar paper on the roofs, and plastic sheeting covered the windows. Could this truly be Tangshan? Could this really be my home? I could hardly believe it. When I finally walked into our own house, the reality was even harder to face. A worn, tattered quilt lay spread on the heated brick bed. There was not a single piece of decent furniture in the room. Even the bowls and plates we used had been given to us by kind neighbors. Looking at my frail, elderly grandmother, my young brother, and bearing the blow of having lost our parents, I could not begin to imagine how we would carry on with the days ahead.

My grandmother was already in her sixties, and providing for my brother and me became a pressing concern. After my father's death, his former workplace gave us a monthly allowance of twenty-two yuan. But our home had been stripped bare. That small sum was nowhere near enough to sustain us. Some suggested that my brother and I be sent to a children's home, which would ease the burden on the family. My grandmother refused. She said firmly, "As long as I am here, you will not go to an orphanage. No matter how hard it is, I will raise you myself."

We truly had nothing. Relatives in Beijing, including my great-uncle, uncles, and aunts, sent money and supplies to help us through the worst of it. Such help, however, could only last for a time. "If we want to solve our problems for good, we have to rely on ourselves," my grandmother would often say.

To make ends meet, I stopped going to school. My father's former workplace agreed to let me take over his job. And so, still in my early teens, I said goodbye to the school I loved, to my teachers and classmates, and gave up the chance to continue my studies. I began working at the very place where my father had once worked. Whenever I saw my former classmates walking to school, something inside me tightened. For our survival, and to ensure that my younger brother could remain in school, I made up my mind to work hard. I was the older sister, after all.

When I first began working, I struggled to adjust to the rotating three-shift schedule. To keep me from nodding off during the night shift, my grandmother did everything she could to make sure I was well rested. In the summer, the makeshift shelter was as hot as a steamer. She would sit beside me and fan me for hours so I could sleep properly. With enough rest, I was able to stay alert at work, and my efficiency improved. The senior workers often praised me, saying that although I was young, I worked no less capably than any adult. As the years passed, my brother and I both married and started families of our own. By then, our grandmother was very old. She could no longer care for herself and needed constant attention. After much discussion, and with great reluctance, we arranged for her to move into a nursing home not far from our house, where there was round-the-clock care. From then on, my brother travelled back and forth between our home and the nursing home every day, year after year, in all kinds of weather. In old age, what a person longs for most is to have family close at hand. Yet we had no better way to care for her. Sending her to the nursing home was a choice born of necessity. All we could do was visit her as often as possible so she would not feel lonely. I just hope she never blamed us.

Our home was farther from the nursing home, so I could not visit her every day. I could only go on days off or whenever I managed to find the time. For that, I have always felt deeply ashamed. Compared with my brother, I fell far short. I admire him from the bottom of my heart. He has always been a devoted and filial son, and everyone recognises it. Respecting and caring for the elderly is a long-standing virtue in our culture. Our grandmother's life was full of hardship. She endured so much to raise my brother and me. Now that she is old, it is our duty as her children and grandchildren to take good care of her. As long as she is still with us, we consider it a blessing.

Nearly forty years have passed in the blink of an eye. I have grown from an unknowing child into a mother myself, and I have grown up alongside the new Tangshan. During these decades, I have witnessed firsthand the city's extraordinary transformation. With strong support from people across the country and the combined efforts of Tangshan's own residents, the local economy has developed rapidly and living standards have risen markedly. In recent years in particular, the city's infrastructure has changed from year to year. Roads now run north to south and east to west, like arteries carrying life through the city, making daily travel far easier. A network of waterways circles the urban area like a jade belt, lending fresh vitality to this reborn city.

Standing atop Fenghuang Mountain and looking out over the city, I can hardly believe that the Tangshan of tar paper roofs weighed down with bricks is gone. In its place rise tall buildings, standing firm and orderly against the skyline. At night, brilliant lights illuminate the streets, dressing the city in colour and glow.

My own family has shared in this change. We left behind the low, makeshift shelter and moved into a spacious apartment, living a life free from want. Household appliances are now commonplace. From our windows, we see neon lights, broad avenues, and bright street lamps. Even at eight or nine in the evening, the shops and supermarkets along the streets remain open. A simple evening walk often turns into a stroll through a lively marketplace. All of this fills the people of Tangshan with quiet pride.

As a survivor of the earthquake and a witness to the rebirth of this city, I carry deep gratitude in my heart for the life we have today. I will always remember those who once helped, rescued, and supported us, and I wish them peace and happiness throughout their lives. I also wish for my hometown to grow ever more beautiful in the years to come.

Chapter Five

Guidelines for Earthquak Preparedness and Disaster Mitigation

5.1 Earthquake and disaster preparedness education

The purpose of this education is to raise awareness of disaster preparedness and mitigation, provide basic knowledge of natural hazards, their impacts, and response strategies, and develop practical skills in emergency evacuation and post-disaster first aid.

5.1.2 Key components of earthquake and disaster preparedness education

a. Understanding earthquake risks
Provide basic knowledge about how earthquakes occur, local seismic zones and geological conditions, possible warning signs, and appropriate protective actions. Residents should also know the locations of nearby emergency shelters and the safest routes to reach them, including alternative routes.

b. Self-rescue and mutual aid skills
Teach essential safety skills for different situations such as at home, outdoors, or at school. This includes how to protect oneself during an earthquake, how to evacuate safely afterward, and how people can assist each other and share practical experience in handling emergencies.

c. Government disaster policies and assistance
Introduce government policies, laws, and support systems related to disaster prevention, mitigation, and relief. Residents should understand how to seek help from authorities, including knowing emergency contact numbers and available assistance channels.

5.1.3 Channels for earthquake and disaster preparedness education

Earthquake and disaster preparedness education can be delivered through community activities, public awareness campaigns, and media outreach. Communities can organise awareness weeks, lectures, film screenings, reading programs, and experience-sharing sessions to help residents learn risk prevention, home safety checks, and basic self-rescue and mutual-aid skills. Public campaigns may use bulletin boards, posters, banners, loudspeakers, and informational leaflets to widely disseminate preparedness knowledge and encourage family participation. At the same time, cooperation with television, radio, newspapers, websites, and social media platforms can help build a broad public communication network through videos, audio programs, and other accessible content.

Comparison of Public Awareness Methods

Awareness Method	Advantages	Disadvantages	Appropriate Use
Training	Memorable and effective	Relatively high cost	Developing disaster preparedness leaders/personnel
Radio Broadcasting	Low cost and wide coverage	Information may not be easily remembered	Announcing latest disaster-related updates
Video / Audio Media	Vivid and engaging	Higher requirements for equipment and content	Delivering disaster knowledge effectively
Brochures and Posters	Easy to carry and distribute	Relatively high cost	Household awareness and preparedness promotion
Word-of-Mouth Communication	Convenient, flexible, and low cost	Requires skilled communicators	Routine disaster awareness activities
Mobile Text Messages	Convenient and wide coverage	Requires access to devices and users	Mass notification of urgent disaster information
Slogans and Signs	Low cost, easy to remember, easy to spread	May become merely symbolic	Mass notification of urgent disaster information

5.2 Emergency response planning and emergency drills

5.2.1 Purpose of regular emergency response planning and drills

Regular emergency planning and drills are conducted to test and improve emergency response plans and overall preparedness. Through these drills, weaknesses in existing plans can be identified and corrected so that the plans become more practical and workable. Regular drills also help assess whether emergency teams, supplies, equipment, and technical support are adequately prepared, allowing any gaps to be addressed in advance. In addition, they help organisations and personnel become familiar with emergency procedures and strengthen their response capabilities. They also clarify the responsibilities of different agencies and improve coordination mechanisms. At the same time, regular drills help raise public awareness of disaster risks and improve people's ability to carry out basic self-rescue and mutual aid during emergencies such as earthquakes.

5.2.2 How to develop an emergency response plan

Developing an emergency response plan generally involves five key steps:
Step 1: Collect and organise relevant information about the community.
Conduct field interviews, questionnaires, and other methods to gain a preliminary understanding of the area's earthquake history, as well as residents' awareness of earthquake preparedness and disaster reduction.
Step 2: Create maps showing earthquake risk and emergency resources.
These may include a hazard distribution map, maps of vulnerable populations and housing, and maps showing the locations of emergency supplies and evacuation or temporary shelter sites.
Step 3: Establish an earthquake emergency response organisation.
This typically includes an Earthquake Disaster Response Command Centre and several working groups, such as a damage inspection group, evacuation and resettlement group, logistics and supply group, and medical assistance group.
Step 4: Classify levels of earthquake emergency response and define activation criteria.
Different response levels should be established according to the severity of the disaster, along with clear standards for activating each level.
Step 5: Define responsibilities for earthquake emergency response.
When an emergency response is activated, the Earthquake Disaster Response Command Centre coordinates and directs the work of the damage inspection, evacuation and resettlement, logistics support, and medical assistance teams.

5.2.3 Preparing an emergency drill script

Emergency drills usually require a drill script. The script is the core document that guides how the exercise is carried out. It can take the form of a narrative document or a structured table, and typically includes an introduction to the drill background, the scenario, narration, and the main activities to be carried out during the exercise.

When preparing a drill script, several points should be kept in mind:
1. The script must be consistent with the emergency response plan.
The drill script should be developed on the basis of the existing emergency response plan. This means gaining a clear and comprehensive understanding of key elements in the plan, such as local geological conditions, disaster risks, emergency response organisations, personnel responsibilities, and response procedures. It is also important to conduct field investigations of potential hazard locations within the community. The scenario should be designed

around the most likely earthquake magnitude for the area, ensuring that the drill is scientifically sound and evidence-based.

2. The script should reflect the local community.

During script development, it is important to communicate with local residents and gather their opinions and suggestions. The script should be revised and improved based on this feedback so that it reflects local realities. A script developed in this way will be more realistic and locally relevant, with more natural dialogue and scenarios. It also helps participants better understand their roles and responsibilities during the exercise.

3. The script should match local conditions.

While the overall structure of the drill can follow a generally applicable framework, the details should incorporate local characteristics and practical conditions. It is advisable not to include excessive dialogue. Instead, organisers should provide face-to-face explanations and guidance to participants so they clearly understand each scenario and can naturally respond to questions such as "What should I do?" and "What should I say?". This approach helps ensure the exercise is realistic, practical, and adaptable for wider use.

5.2.4 Evaluation of emergency drills

Evaluation is an important step after an emergency drill. Once the exercise is completed, the organising body should bring together relevant personnel to review and analyse the issues identified during the drill, and assess whether the exercise achieved its intended objectives. The purpose is to improve disaster preparedness and strengthen participants' emergency response skills.

The main steps in evaluating a drill include:

1. Observing and recording the activities during the exercise
2. Comparing participants' performance with the objectives and requirements of the drill
3. Summarising and organising the problems identified during the exercise
4. Putting forward recommendations for improvement

Common methods used in drill evaluation include:

1. Review and interviews conducted by evaluators
2. Reports and self-assessments from participants
3. Open discussions or meetings to exchange views and reach consensus

5.3 First aid training for earthquake emergencies

Accidents, disasters, and sudden injuries often occur unexpectedly, and in many cases professional medical personnel are not immediately available at the scene. For this reason, providing correct basic first aid on site is extremely important. Early and appropriate assistance can create favourable conditions for later medical treatment and help save valuable time for professional rescue and hospital care.

At the scene of an accident, there are often multiple casualties, including some who may be seriously injured or in critical condition. In such situations, timely, accurate, and appropriate first aid becomes crucial for saving lives. The main purpose of on-site first aid is to ensure that injured individuals receive initial treatment as quickly as possible, stabilising their condition and buying time for further medical treatment once they are transported to a hospital.

5.3.1 Burn injuries

Burn injuries may occur after an earthquake due to fires, hot liquids, or explosions. Providing quick and appropriate first aid can help reduce the severity of injuries and improve survival chances.

Remove the source of the injury immediately and cool the affected area. If the burn is caused by hot liquids, remove any clothing soaked with the liquid at once and rinse or immerse the injured area in cool water as soon as possible. If the burn is caused by flames, the injured person should roll on the ground immediately, or the burning area should be covered with a quilt, blanket, or similar material to extinguish the fire. Where it is safe to do so, use water to put out the flames. If water is not suitable, use a fire extinguisher or other appropriate equipment. After the source of the burn has been removed, the wound should be rinsed with cool water to reduce the temperature of the burned area.

Remove the injured person from the fire or heat source as quickly as possible to reduce the duration of burning. Avoid actions that may worsen the fire. For example, running can make burning clothes flare up, while standing upright may cause hair to catch fire and increase the risk of inhaling smoke, which may lead to burns in the respiratory tract. Instead, the person should lie down immediately and roll slowly on the ground to extinguish the flames.

Check the person's heartbeat and breathing, and determine whether there are other injuries, toxic gas inhalation, or related complications. For victims of explosion-related burns, check carefully for head injuries, chest injuries, internal organ damage in the chest or abdomen, and burns to the airway.

Prevent shock, suffocation, and wound contamination. Burn victims often experience shock due to severe pain or fear, so pain relief medication may be given if available. If acute airway obstruction or suffocation occurs, medical personnel should be contacted immediately so that emergency procedures, such as establishing an airway, can be performed. During examination and transport, care should be taken to protect the wound and prevent infection. Leave the dangerous area quickly and transport seriously burned patients to hospital as soon as possible. During transport, handle the injured person gently and steadily, and continuously monitor their condition.

Remove burned clothing, shoes, and socks as quickly as possible. To save time and minimise further damage to the wound, clothing can be cut away with scissors. Do not attempt to clean the wound at the scene. Instead, cover the burn loosely with clean clothing or cloth to prevent contamination and reduce pain caused by air exposure. Do not break blisters, and avoid applying ointments, oils, or other unproven remedies to the wound.

5.3.2 Traumatic injuries

a. Types of bleeding

Bleeding can be classified according to the type of blood vessel involved:

- Arterial bleeding: The blood is bright red and spurts out in pulses. This type of bleeding is very dangerous and requires immediate attention.
- Venous bleeding: The blood is dark red and flows out more steadily. It is still potentially dangerous.
- Capillary bleeding: The blood is bright red and slowly oozes from the surface of the wound. This type of bleeding is usually less serious.

b. Methods of stopping bleeding

- General method: Clean the wound → disinfect it → apply a bandage.
- Direct pressure with fingers: Use the thumb to press the artery above the wound (towards the heart) to block the blood flow. This method is mainly used for emergency control of arterial bleeding and should not be applied for too long.
- Tourniquet method: Raise the injured limb for about two minutes, place a soft pad under the tourniquet, and then tighten it to stop the bleeding.
- Packing method: Pack the wound with sterile gauze or similar material, apply pressure, and then bandage the wound. This method is commonly

used for venous bleeding.

- Pressure bandaging: This is a direct pressure method and the most commonly used way to control bleeding. If there are no foreign objects or bone fragments in the wound, place a clean dressing over the wound, then use a bandage, triangular cloth, or wide cloth strip to apply firm pressure until the bleeding stops.

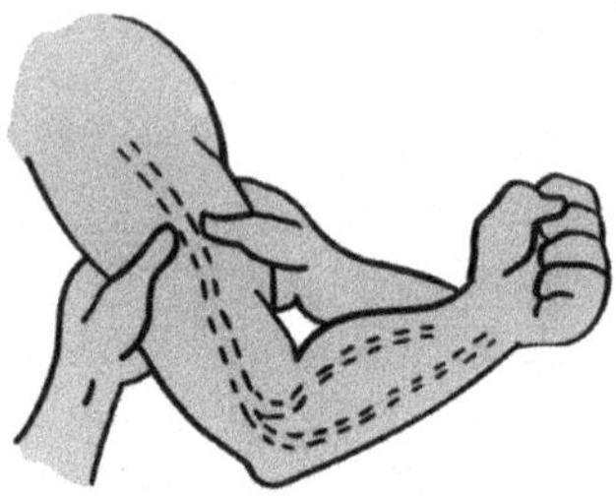

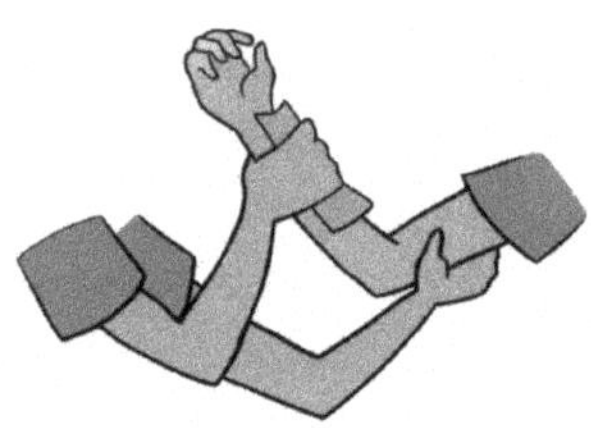

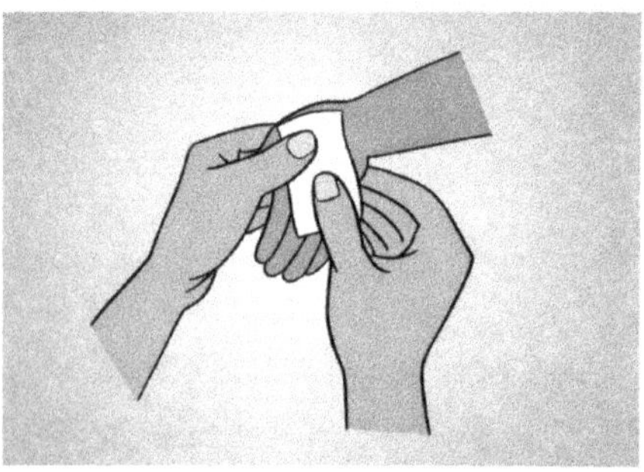

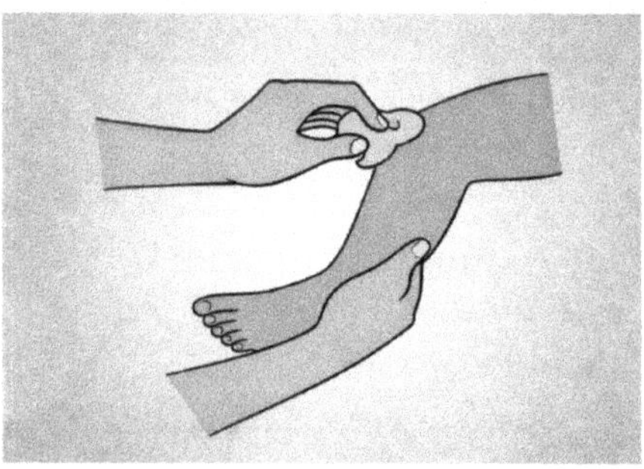

5.3.3 Cardiopulmonary resuscitation (CPR)

a. Purpose of CPR

The brain requires a continuous supply of oxygen and is highly sensitive to oxygen deprivation. When the heart suddenly stops beating, blood circulation stops and oxygen can no longer reach the brain. Without prompt treatment, permanent brain damage may occur within about 4–6 minutes. Cardiopulmonary resuscitation (CPR) is an emergency procedure used to maintain blood circulation and breathing until professional medical help arrives. Starting CPR as early as possible greatly increases the chances of survival.

b. Basic steps of CPR

Modern CPR follows the CAB principle, which stands for Circulation (chest compressions), Airway (opening the airway), and Breathing (rescue breath-

ing).
Chest compressions: place the person on a firm surface. Put the heel of one hand on the centre of the chest (lower half of the sternum) and the other hand on top. Keep your arms straight and press down using your body weight. Perform 100–120 compressions per minute, with a depth of about 5–6 cm for adults, allowing the chest to fully recoil after each compression.

Open the airway: after 30 compressions, open the airway using the head-tilt, chin-lift technique.

Rescue breathing: give two rescue breaths, each lasting about one second, ensuring the chest rises. Continue cycles of 30 compressions and 2 breaths (30:2).

c. Checking for breathing
Breathing can be assessed by listening, looking and feeling. Listening for breathing sounds, looking for movement of the chest, and feeling for airflow against your cheek.

d. When to stop CPR
CPR should continue until one of the following occurs: the person regains normal breathing and a heartbeat; another trained rescuer takes over; medical professionals arrive and assume responsibility; the rescuer becomes physically exhausted and cannot continue; or a doctor declares the person deceased.

5.3.4 Immobilisation of fracture

Fractures are one of the most common types of traumatic injury and are often accompanied by damage to blood vessels. Typical symptoms include local pain, swelling, deformity, and impaired function of the affected area. Fractures are generally classified into closed fractures and open fractures.

The basic principles of first aid for fractures are as follows: if the injured person is in shock, treat the shock first and then immobilise the fracture; for open fractures, stop the bleeding and bandage the wound before immobilising the injured limb.

For patients with fractures, joint injuries, or extensive soft-tissue damage, the injured area should be properly immobilised before transportation. Transport methods include using a stretcher or manual carrying. In particular, patients with suspected spinal injuries must not be moved carelessly, as improper movement may cause further injury.

5.4 Planning and establishing school emergency shelters

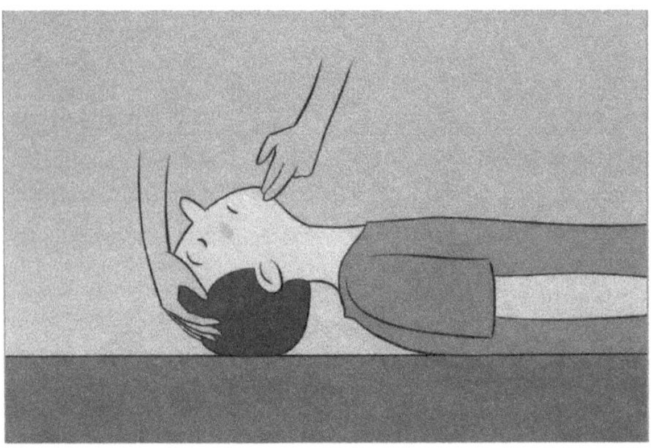

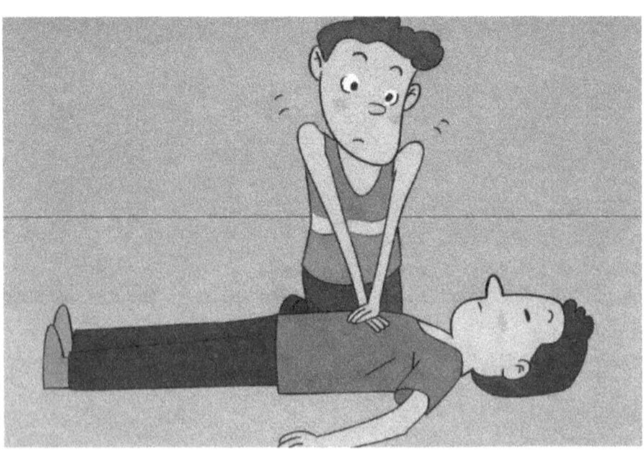

5.4.1 Importance of establishing emergency shelters in schools

Disaster evacuation shelters are places used for evacuating affected people during disasters and providing temporary living space after disasters. These shelters are generally classified into emergency shelters, fixed shelters, and central shelters. Primary and secondary schools have many advantages that make them suitable for use as emergency shelters. They are usually well distributed, well equipped, widely recognised by the public, convenient for daily access, and suitable for evacuation and mutual assistance. Because of these advantages, schools are considered an important resource for emergency shelter planning in both urban and rural areas.

Compared with many other buildings in the same area, schools are often designed with higher earthquake-resistance standards, making them relatively safer during disasters. Schools also typically have large, flat open spaces, such as sports fields, running tracks, ball courts, and green areas, as well as spacious indoor areas, including activity halls or multi-purpose rooms. These spaces make it possible to accommodate a large number of evacuees when necessary.

The location of schools usually corresponds closely with population distribution, since schools are normally established to serve specific communities or villages. In cities, schools are numerous and widely distributed, often located in areas with convenient transportation. Local residents are generally familiar with their locations and access routes. In villages and small towns, schools are often one of the most important public buildings and are well known to local residents.

Many schools are already equipped with basic infrastructure needed for emergency shelters, including communication and broadcasting systems, water supply, sanitation facilities, drainage systems, electricity supply, and fire-fighting equipment. If emergency relief supplies are also stored in advance, schools can quickly be converted into temporary disaster shelters when a disaster occurs.

5.4.2 Selecting the location of shelters

A disaster shelter should be located in an area with sufficient open space, allowing people from nearby communities to gather safely. Suitable locations in schools include green areas, playgrounds, sports fields, or gymnasiums.

Shelters should be placed in safe areas that are not prone to geological hazards, such as landslides or other secondary disasters. The site should be suitable for setting up temporary tents and conducting rescue operations. It should also provide favourable conditions for fire safety, public security, sanitation, and disease prevention, ensuring that evacuees remain safe and healthy.

It is also necessary to clearly identify and record the shelter location, its capacity, and the personnel responsible for its management. This information should be available before disasters occur so that the shelter can be quickly activated and used for emergency relief when needed.

5.4.3 Supporting facilities and functional layout

To function effectively as a disaster shelter, schools should be equipped with basic supporting facilities. These may include open sports areas, storage rooms, water storage tanks, drainage and sewage systems, space reserved for temporary toilets, and clearly marked evacuation routes.

Where conditions allow, additional facilities should also be prepared. These may include temporary accommodation areas for teachers and students, storage for daily necessities and medical supplies, fire-fighting equipment, emergency communication and broadcasting systems, temporary power generation and lighting equipment, and basic medical facilities.

Emergency shelters should ideally have at least two entrances and exits to allow safe and efficient movement of people. Whenever possible, entrances for vehicles and pedestrians should be separated to avoid congestion. The design of shelters should also consider the needs of different groups. Facilities should accommodate adults and children, and special attention should be given to elderly people, persons with disabilities, and other vulnerable groups by incorporating appropriate accessibility features.

5.5.4 Preparation and inspection of emergency supplies

Emergency shelters should maintain a reserve of essential relief supplies. These supplies typically include rescue tools such as shovels, stretchers, and fire extinguishers; communication equipment such as loudspeakers and walkie-talkies; lighting equipment including flashlights and emergency lamps; basic medical supplies such as bandages and common medicines; and daily living items such as blankets, tents, sleeping bags, food, and drinking water. Additional emergency equipment may include fire-fighting tools and facilities, warning signs and symbols, respirators, and gas masks. Medical supplies should also include medicines needed for treating injuries such as burns, poisoning, or other emergency conditions.

School disaster management personnel should conduct regular inspections of all emergency supplies and equipment. These inspections should check storage conditions, functionality, and completeness of the equipment. If any items are missing, damaged, or malfunctioning, they should be replaced or repaired immediately.

A designated person should be responsible for routine inspections. Electrical equipment should undergo regular charging and discharging, and communication devices such as telephones should be tested to ensure they are functioning properly. Fire extinguishers should be inspected regularly to check for external damage and operational readiness. Their weight should be checked at least every six months, and if the weight falls below the required level, they should be serviced and refilled by qualified maintenance personnel.

Protective equipment such as gas masks should also be inspected regularly and kept clean and dry to prevent corrosion or damage. A comprehensive inspection should be carried out at least once every two weeks, and any defective equipment should be replaced as soon as possible.

Through proper planning, facility preparation, and regular inspection, schools can become safe and effective disaster shelters that support evacuation, relief operations, and temporary living needs after disasters.

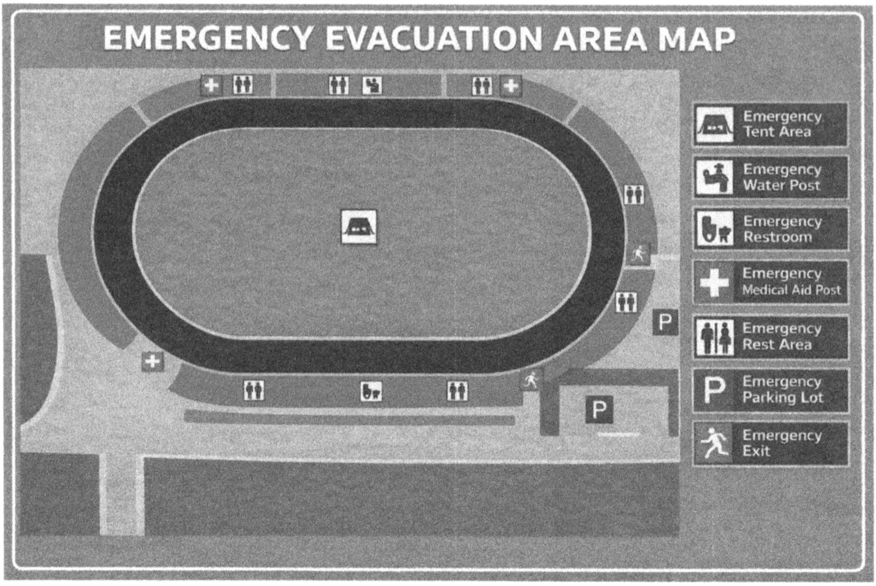

Post-Earthquake Fire Safety Knowledge Survey

No.	Survey Question	Knowledge Type
1	If a fire breaks out at school, how should you escape?	Emergency Knowledge
2	Do you know how to use a fire extinguisher?	Response Behavior
3	Can you operate a fire hydrant?	Response Behavior
4	What is the correct way to use a fire extinguisher?	Emergency Knowledge
5	Do you know where fire extinguishers are located in your school?	Preparedness Behavior
6	If a fire occurs and thick smoke fills the room, what should you do?	Response Behavior
7	When staying in a hotel during travel, what should you check first?	Prevention Knowledge
8	What is the emergency phone number to report a fire?	Emergency Knowledge

Chapter Six

Understanding Earthquakes

6.1 What is an earthquake

An earthquake, also known as a seismic event or ground tremor, is a natural phenomenon caused by the rapid release of energy within the Earth's crust, resulting in vibrations that generate seismic waves.

6.2 Magnitude and intensity of earthquakes

Earthquake magnitude refers to the size of an earthquake, determined by the amount of energy released during each seismic event as measured by seismographs.

Earthquake intensity refers to the degree of damage caused by an earthquake to the ground and to structures such as buildings. Areas closer to the hypocenter (earthquake source) suffer greater damage and higher intensity, while areas further away experience less damage and lower intensity.

Factors affecting earthquake intensity include: (1) earthquake magnitude; (2) hypocenter depth; (3) distance from the epicenter; (4) soil and geological conditions; (5) building performance; (6) focal mechanism; and (7) topography and groundwater level.

The China Seismic Intensity Scale classifies earthquake intensity into different levels based on human perception and the degree of earthquake damage. At present, there are many types of seismic intensity scales, with the 12-degree scale being the most common. There are also 7-degree scales (used in Japan) and 10-degree scales. China adopts the 12-degree seismic intensity scale.

China Seismic Intensity Scale

Intensity	Human perception on the ground	Degree of building damage	Other damages
I	Imperceptible		
II	Felt by a few people at rest indoors		
III	Felt by a small number of people at rest indoors	Slight rattle of doors and windows	Slight sway of hanging objects
IV	Felt by most people indoors and a few outdoors; some awakened from sleep	Rattle of doors and windows	Hanging objects sway noticeably; utensils rattle
V	Fely by most people indoors and outdoors; many people awakened from sleep	Doors, windows, roofs and roof frames shake; plaster falls; fine cracks appear; roof tiles may fall; in isolated cases, chimneys collapse	Unstable objects sway or topple
VI	Most people have difficulty standing; some flee outdoors	Damage occurs, wall cracks appear; roof tiles fall; a few chimneys crack or collapse	Cracks appear on riverbanks and in loose soil; saturated sand layers may exhibit sand boiling and water sprouting; some freestanding brick chimneys crack severely
VII	Most people flee outdoors; cyclists feel the shaking; passengers in moving vehicles feel it	Light damage - partial collapse; cracks appear; minor or no repairs needed	Riverbanks collapses; saturated sand layers often exhibit sand boiling and water sprouting; more cracks in loose soil; most freestanding brick chimneys suffer moderate damage
VIII	Most people are strongly shaken and have difficulty walking	Moderate damage - structural damage; repairs needed prior to continued use	Hard and compacted soil cracks; most freestanding brick chimneys damage severely; tree tops break; building damage results in casualties among people and animals
IX	People in motion fall	Severe damage – severe structural damage with partial collapses; difficult to repair	Hard and compacted soil cracks a lot; bedrock may have fractures or displacements; landslides and collapses are common; freestanding brick chimneys collapse
X	Cyclists fall; people in an unstable state be thrown off their feet, experiencing a sense of being lifted	Most buildings collapse	Landslides and earthquake-induced faults occur; arch bridges on bedrock are damaged; most freestanding brick chimneys are destroyed or broken off at the base
XI		Almost everything collapses	Earthquake faults extend over long distances, accompanied by numerous landslides and slope failures
XII			The ground undergoes dramatic changes, with mountains and rivers transformed

6.3 How earthquakes occur

An earthquake is a natural phenomenon in which ground shaking is caused by movements within the Earth. On Earth, tectonic plates continuously interact with one another. The compression and collision between plates cause faulting and fracturing along plate boundaries and within plates, which is the primary cause of ground vibration, that is, earthquakes.

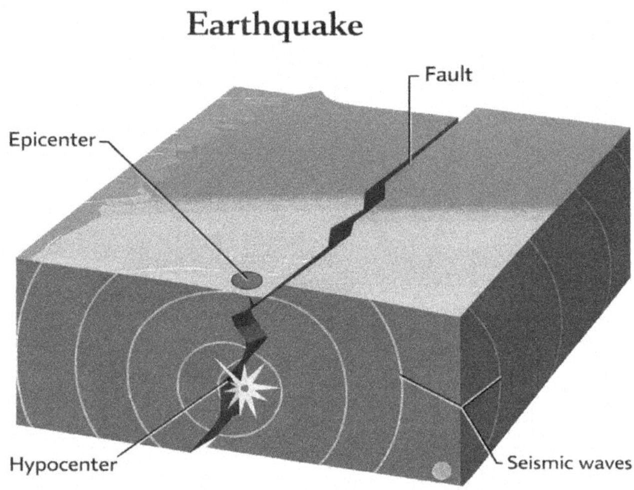

6.4 Earthquake location

To determine the earthquake location means we need to acquire the basic information of the position of the epicenter, the depth of the hypocenter, the time when the earthquake occurred and the magnitude of the earthquake.

Several methods are commonly used to locate an earthquake. There are Ishikawa method, Wadati method and Takahashi method. Azimuth method and Intersection (Triangulation) method are used for a quick report. However, these two methods are less accurate. More precise methods, such as Master-event location method and the joint hypocenter determination method, can provide very accurate results, but they require strict conditions and are therefore mainly used in specialised research.

The figure below shows the intersection method.

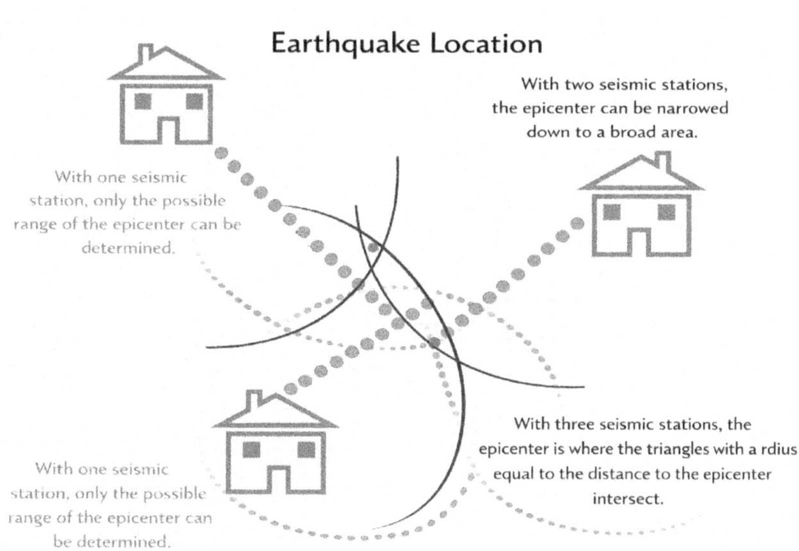

6.5 Earthquake terminologies

Earthquake naming: earthquakes are named after the county where the epicenter is identified by instruments.

Hypocenter: hypocenter is the place inside the earth where an earthquake begins. It usually refers to the point underground where rocks first break and shift, releasing energy and generating seismic waves.

Focal depth: if the hypocenter is considered as a single point, the vertical distance from this point to the ground is the focal depth.

Epic-enter: The epicenter is the point on the ground that lies directly above the hypocenter.

Epicentre distance: the epicentre distance is the distance on the ground from the epicentre to a specified point.

Isoseismal line: An isoseismal line is a contour line that connects points with the same earthquake intensity.

Meizoseismal area: the meizoseismal area is the region that suffers the most severe damage during an earthquake. The geometric center of this area is called the macroscopic epicentre.

Earthquake swarm: an earthquake swarm is a sequence of earthquake occurring within a certain period of time in the same hypocenter region. These earthquakes vary in size but are related, sharing similar

causes or geological structures. Earthquakes can be categorised into three types, which are main-shock type, swarm type and isolated type.
Main-shock: the main-shock is the earthquake with the largest magnitude in an earthquake sequence.
Aftershock: An aftershock is a smaller earthquake that occurs after the main-shock in the same region.
Fore-shock: A fore-shock is a smaller earthquake that occurs before the main-shock in the same region.

6.6 Seismic zones

A seismic zone or a seismic belt is a region where earthquakes occur frequently and tend to focus. There are three major seismic belts in the world: the Pacific Ring of Fire, the Eurasian Seismic Belt, and the Mid-Ocean Ridge Seismic Belt. China is located in the south-eastern Eurasian continent, between the Pacific Ring of Fire and the Eurasian Seismic Belt.

Global Map of Volcanic and Seismic Belts

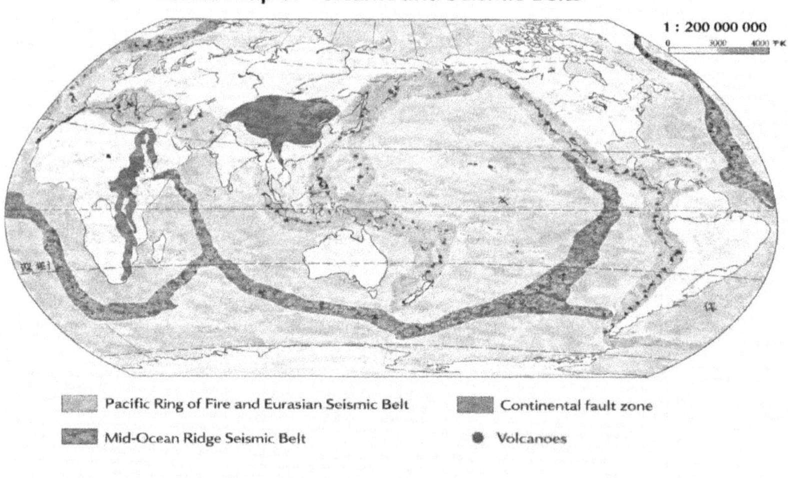

6.7 Earthquake predictions in China

At present, scientists cannot accurately predict most destructive earthquakes. However, by making full and reasonable use of existing practical experience and scientific research findings, it may be possible to make limited predictions for some specific types of earthquakes.

China, like other countries around the world, is still in the exploratory stage of earthquake prediction, and the overall level of prediction remains limited. Compared with countries such as Japan and the United States, China still lags behind in the advancement of observation technologies and in the theoretical foundations of earthquake prediction.

However, China has a clear advantage in the abundance of earthquake case data and the accumulation of on-site prediction experience. The frequent occurrence of moderate to strong earthquakes in China has provided earthquake scientists with many opportunities to carry out practical prediction experiments.

In the mid-1970s, China successfully predicted several destructive earthquakes, most notably the 1975 Haicheng earthquake (magnitude 7.3) in Liaoning Province. Since the 1990s, China has also achieved a series of successful short-term and imminent earthquake predictions, including the 1995 Menglian earthquake (M7.3) in Yunnan, the 1997 Jiashi earthquake sequence in Xinjiang (M6.3 and M6.4), four earthquakes of magnitude 5.0 or above in Ninglang County, Lijiang, Yunnan in November 1998 (M6.2), and the December 29, 1999 Xiuyan–Haicheng earthquake (M5.6) in Liaoning. Such a number of successful short-term earthquake predictions is rare on a global scale. Overall, China's earthquake prediction capability is considered to be among the more advanced levels in the world.

6.8 Fault activity and geological hazards in Tangshan

The urban area of Tangshan is subject to frequent earthquakes, many of which cause serious damage, have wide impacts, and occur in sequences. The region is also affected by geological hazards such as karst-related ground collapses and ground subsidence in mined-out areas.

The 1976 Tangshan earthquake (magnitude 7.8) caused extremely serious damage. More than 240,000 people lost their lives, over 4.25 million buildings were destroyed, and much of the city was severely damaged. Large coal mines, such as the Kailuan coal mine, were flooded and forced to suspend operations, resulting in enormous economic losses. This event is recognised as one of the most devastating earthquakes of the 20th century.

The severity of the disaster was closely related to the complex geological structure of the Tangshan area and its frequent seismic activity.

6.8.1 Geological tectonic activity background

The location of Tangshan City is at a junction of a structural belt and a tectonic framework. The region has numerous active faults, many of which are large in scale. A series of north–northeast (NNE)-trending faults and fold belts have developed. From east to west, these structures include the Kaiping (开平) syncline[5], the Beziyuan (碑子院) anticline, and the Chezhoushan (车轴山) syncline. (see the regional structural geology sketch map).

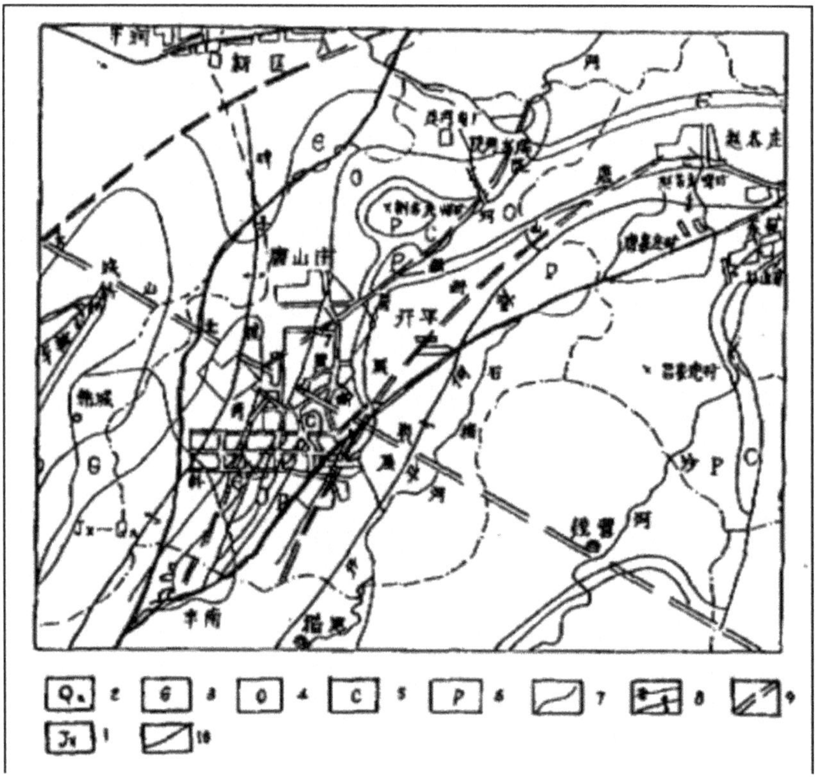

[5] *The fold structures passing through the urban area include the Kaiping Syncline, which is approximately 50 km long and 20 km wide, and the Beziyuan Anticline, which has a total length of about 30 km and a width of 10 km.*

6.8.2 Fault structures

Several important fault systems pass through or near the Tangshan urban area.
(1) Douhe Fault Zone: the Douhe Fault Zone consists of two main branches, extending roughly in an east–west direction.
(2) Tangshan Mine–Zhaogezhuang Fault Zone: this fault zone is composed of several discontinuous fault segments and extends for more than 30 km, trending generally northeast.
Tangshan Mine Fault: this fault lies in the south-western part of the fault zone and extends through the south-eastern urban area of Tangshan.
Western Zhaogezhuang Mine Fault: located at the north-eastern end of the fault zone and farther away from the city, this fault shows clear evidence of both vertical and horizontal movement, indicating complex fault motion.
(3) Shexian–Lulong Deep Fault: this fault is 2–4 km wide and is considered an ancient, concealed fault buried beneath the surface.
(4) Dachengshan North Slope Fault: this fault trends roughly west–northwest and extends for about 60 km. It shows evidence of right-lateral horizontal movement. South of this fault lies a broad uplift zone formed by folds and gentle ground deformation. Geological hazards along this fault are generally less severe than in nearby areas. This structure acted as a barrier fault during the 1976 Tangshan earthquake, influencing the rupture process.

Overall, the faults crossing the Tangshan urban area are among the most active faults in North China and continue to undergo slow and uneven movement to this day.

6.8.3 Earthquake-related geological hazards

The 1976 Tangshan earthquake (magnitude 7.8) occurred on 28 July, with its epicenter located on Jixiang Road in Lunan District. In most parts of the city, the earthquake intensity reached Level XI, causing extremely serious destruction. Only a few areas experienced slightly lower intensities (Level IX–X), mainly due to more favourable local ground conditions. During the earthquake, a wide range of geological hazards occurred in the urban area, including surface faulting, ground fissures, soil liquefaction, riverbank sliding, and damage caused by strong ground motion.

(1) Ground fissures and fault rupture
After the earthquake, studies confirmed that the seismic source structure was the Tangshan Mine Fault No. V. The surface rupture zone extended for about

10 km, generally trending northeast, and passed through several parts of the city.

(2) Riverbank sliding damage

Strong shaking caused soil liquefaction, which greatly reduced soil strength. As a result, large sections of the Douhe River banks slid toward the river channel. This movement led to ground cracking behind the sliding zones, collapse and subsidence of embankments, riverbed uplift, and sediment accumulation.

In some sections, riverbank sliding caused the river channel to narrow, leading to local flooding. Bridges suffered serious damage, including tilting of piers, displacement, and collapse. Roads along the riverbanks were destroyed, and pipelines crossing the river were severely deformed. In several areas, landslides from both banks pushed toward the center of the river, blocking water flow, causing overflow, and flooding large areas of farmland. Pumping stations and flood-control facilities along the river were also damaged.

Chapter Seven

Pre-Earthquake Phenomena

7.1 Natural environment

Seen from space, there is a beautiful blue planet—this is our home, the Earth. From the moment it was formed, the Earth has never stopped moving and changing. The Earth is composed of four interconnected spheres: the atmosphere, the hydrosphere, the biosphere, and the lithosphere.

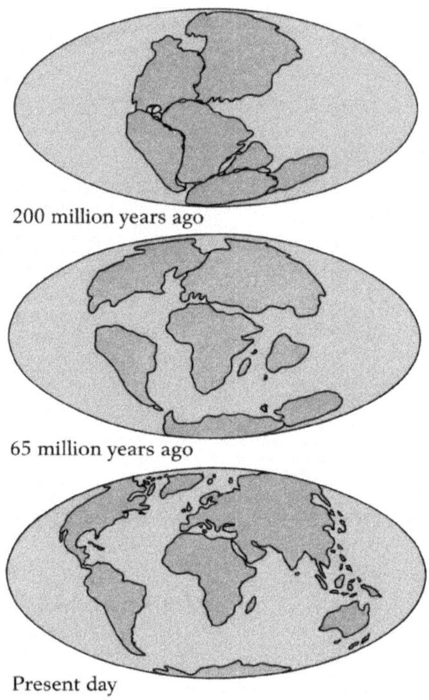

The atmosphere is the outermost gaseous layer of the Earth. The hydrosphere includes oceans, rivers, lakes, wetlands, glaciers, and groundwater. With the presence of the atmosphere, the hydrosphere, and surface minerals, and under suitable temperature conditions, the Earth has developed a natural environment that supports life. This life-supporting system is known as the biosphere.

The outermost layer of the Earth is called the crust, the layer beneath it is the mantle, and the central part of the Earth is known as the core. The Earth has an average radius of about 6,370 kilometers, while the thickness of the crust is only about 35 kilometers. Most destructive earthquakes occur within the Earth's crust. If the mantle and the core are compared to an egg white and an egg yolk, then the Earth's crust is like the eggshell.

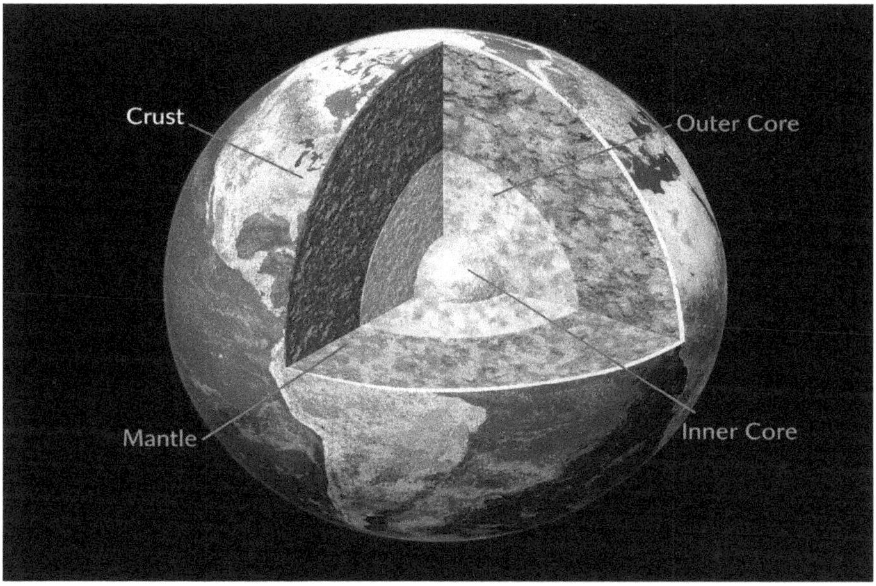

The Earth's surface is not a single rock layer. Instead, it is made up of tectonic plates of different sizes that fit together like pieces of a puzzle. There are seven major plates: the Antarctic Plate, Eurasian Plate, North American Plate, South American Plate, Pacific Plate, Indo-Australian Plate, and African Plate. These plates float on the mantle and move at speeds of a few centimetres to more than ten centimetres per year, gradually drifting, squeezing against one another, and colliding.

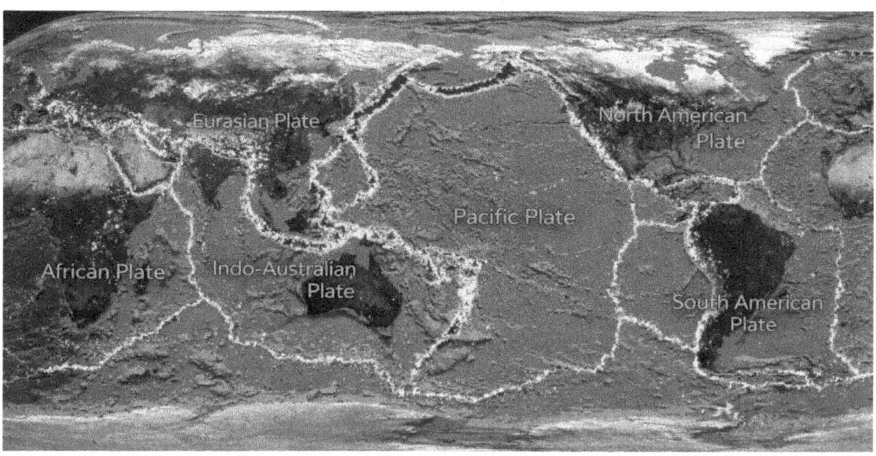

7.1.1 Theory and hypothesis

Continental Drift

The seven major tectonic plates float on a softer layer beneath them, much like ice floating on water. These plates move very slowly, only a few centimetres each year. All continents sit on top of these plates. When the plates move, the continents move along with them. As a result, the positions of continents change slowly over long periods of time. This slow movement of continents is known as continental drift.

Seafloor Spreading

The seafloor spreading theory explains how new ocean floor is formed. Along the middle of the oceans, there are long cracks or rift zones that cut through the Earth's outer layers. Hot material from deep inside the Earth rises up through these cracks, cools down, and forms new oceanic crust. As more material rises and solidifies, the newly formed seafloor pushes older oceanic crust outward on both sides. In this way, the ocean floor is continuously created and renewed. Seafloor spreading provides important evidence for continental drift and represents a further development of the continental drift theory.

7.2 Geological hazards

Geological hazards generally refer to disasters caused by geological processes that result in loss of life and property. There are more than 30 different types of geological hazards. Geological hazards triggered by natural factors, such as rainfall, snowmelt, and earthquakes, are called natural geological hazards. Those caused by human activities, including excavation, loading, blasting, and waste dumping, are known as human-induced geological hazards. The most common geological hazards are those that pose serious threats to human safety and property, such as rockfalls, landslides, debris flows, ground fissures, ground collapse, and land subsidence. All of these hazards are closely related to geological processes.

7.2.1 Rockfall

A rockfall refers to a geological process in which rock or soil on a steep slope suddenly breaks away under the force of gravity. The material then falls, rolls, and accumulates at the foot of the slope or in a valley. Rockfalls usually happen suddenly, move very fast, and can cause serious damage. After a rockfall, piles of fallen rocks and debris are often left behind. This type of disaster is also known as collapse, cave-in, or slope failure.

Main triggers of rockfalls:

1. Earthquakes: earthquake shaking can disturb the stability of a slope and cause rocks to fall.
2. Water: melting snow and rainfall, especially continuous heavy rain, allow water to seep into the slope. This weakens the rock structure and reduces support, making collapse more likely.
3. Human activities: activities such as excavation, mining, reservoir water storage, and flood discharge can damage the original balance of slopes and trigger rockfalls.
4. Terrain and landform conditions: Rockfalls often occur on riverbanks, lake-shores, mountain slopes, as well as along railways and roads. Special attention should be paid to rocks that appear unstable or ready to fall.

Warning signs before a rockfall:

1. Cracks at the collapse site gradually widen, and small pieces of rock may fall from the front edge of unstable rock masses. Minor rockfalls and small collapses may occur repeatedly.
2. New cracks appear at the top of the slope, and in some cases, unusual odors may be noticed.
3. Sounds of rock tearing, rubbing, breaking, or falling may occasionally be heard. At the same time, abnormal changes may appear, such as unusual heat, increased radon gas, or changes in groundwater quality or water volume.

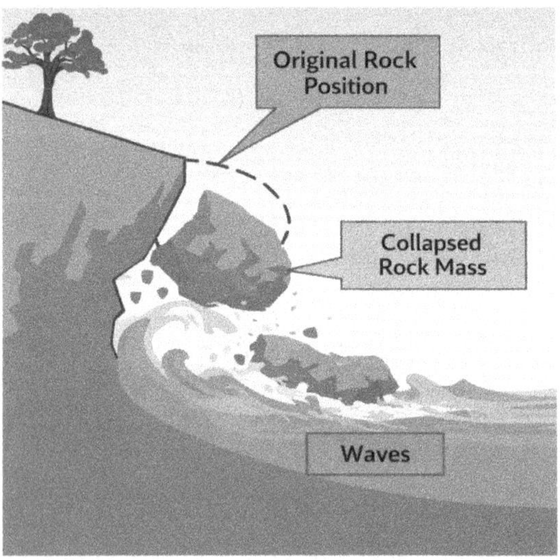

7.2.2 Landslide

A landslide is a natural phenomenon in which rock or soil on a hillside slides downhill under gravity along a weak surface or weak zone. The movement may occur as a whole or in scattered pieces. When a landslide happens, the ground cracks and the hillside collapses, threatening the lives of people and animals. Houses on the slope may collapse, trees may fall, and farmland can be seriously damaged or buried by sliding rocks and soil, causing heavy property losses. Landslides usually occur on slopes with an angle of less than 50 degrees. They are commonly triggered by heavy rainfall, river erosion, earthquakes, and melting snow. In rural areas, landslides are also known by local names such as "ground sliding," "mountain movement," or "mountain collapse."

Main triggers of landslides:
Earthquakes, rainfall, melting ice and snow, and human activities can all trigger landslides. As a result, landslides often occur in clusters and are most common during periods of continuous heavy rain. In south-western China, landslides, debris flows, and rockfalls mainly occur from June to September, while in north-western China, they most often occur in July and August.

During an earthquake, the ground shakes violently, making hillsides very unstable. This shaking can easily cause slopes to collapse and slide downward, leading to landslides. This is because earthquakes release a huge amount of energy, many of the large-scale landslides around the world are caused by earthquakes.

Warning signs before a landslide:

1. Collapses occur at the front edge of the slope, and clear cracks form. The collapse boundary gradually moves upslope, while cracks become longer, wider, and more numerous. In particular, when wide cracks appear at the top of the slope and the cracks are clearly offset, a landslide is very likely to happen.
2. Mound-like bulges appear at the lower part of the slope, and fan-shaped or radial cracks open at the top.
3. Local subsidence occurs on the slope, creating an uneven surface with many pits and depressions.
4. Buildings on the slope deform, crack, or tilt.
5. Water and spring water become muddy, water levels change suddenly, and water may seep out of places that were previously dry.

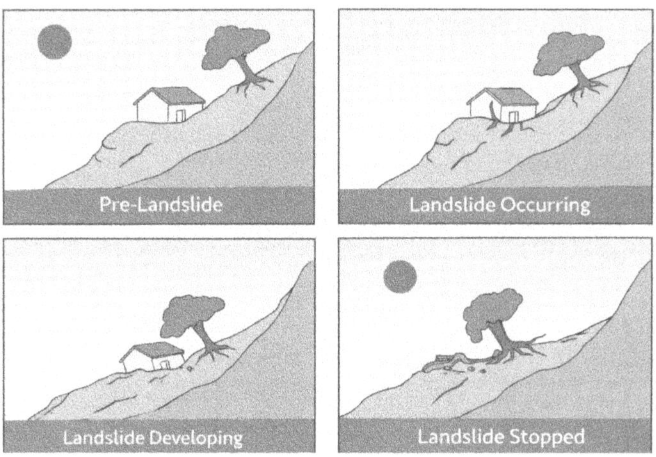

7.2.3 Debris flow

A debris flow is a special type of flood that occurs in mountain areas or deep valleys with steep terrain. It is usually triggered by heavy rain, heavy snow, or other natural disasters, which cause landslides that mix large amounts of mud, sand, and rocks into a fast-moving flow. Debris flows are characterised by sudden occurrence, high speed, large volume, and strong destructive power.

Main triggers of debris flows:
Earthquakes and volcanic eruptions can destroy the stable structure of rock and soil on mountain slopes. Once the loosened rocks and soil are washed by heavy rain or melting snow and ice, large-scale debris flows can easily form.

Human activities, such as soil excavation, quarrying, road construction in mountainous areas, and deforestation can also weaken slope stability. During the rainy season, these activities often lead to debris flow disasters.

Warning signs before a debris flow:

1. Sudden interruption of stream flow or a rapid increase in water volume.
2. Valleys produce loud roaring sounds or with slight ground vibrations.
3. Landslides or collapses along valley sides block channels and supply large amounts of loose material.

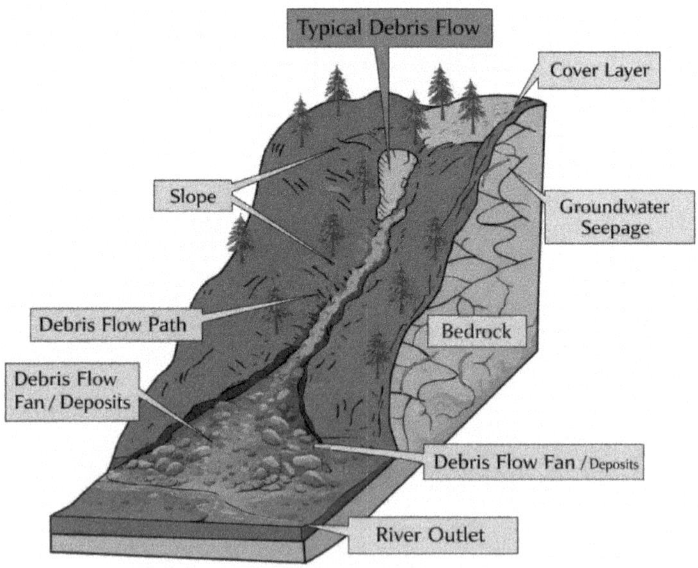

7.3 Pre-earthquake warning signs

Before an earthquake, various natural phenomena may appear that are related to the occurrence of an earthquake. Common pre-earthquake phenomena include abnormal changes in groundwater, unusual animal behaviour, crustal deformation, increased small earthquake activity, as well as changes in ground sounds, ground lights, the Earth's magnetic field, gravity, ground electricity, ground stress, and the concentration of radon or other chemical components in groundwater.

Earthquake precursor anomalies can be divided into microscopic anomalies and macroscopic anomalies. Microscopic anomalies, such as ground deformation and electromagnetic changes, are mainly detected using high-precision scientific instruments. Macroscopic anomalies are those that can be felt by people directly.

7.4 Changes in well water

Changes in groundwater level, as well as changes in water colour, taste, bubbling, foaming, or sudden changes in hot spring temperature, may all be possible earthquake precursors. However, groundwater changes are complex and must be carefully analysed. It is important to exclude the influence of rainfall or snowfall, drought, river level changes, snow and ice melting, human pumping and irrigation, and environmental pollution before relating such changes to earthquakes.

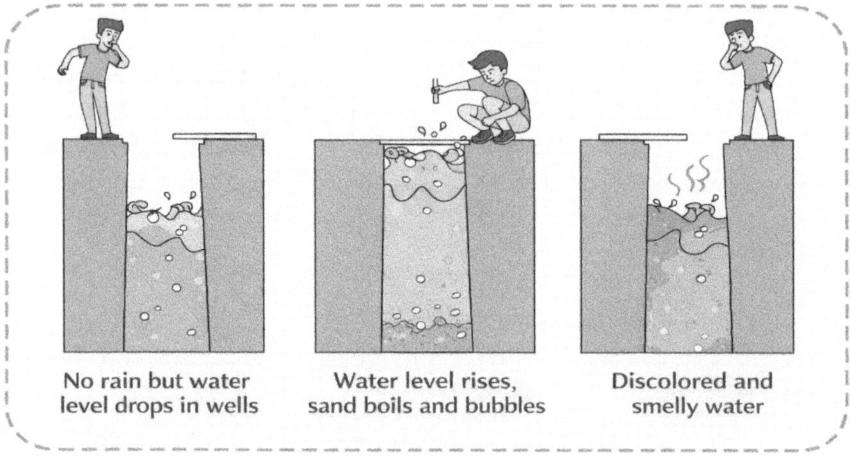

7.5 Abnormal animal behaviours

Historically, animals have shown different kinds of abnormal behaviours before many major earthquakes. Because of this, people often call animals "living instruments" for observing earthquake precursors.

Over time, these observations were turned into folk sayings to help people remember the warning signs. Such sayings describe animals behaving in unusual ways, for example:

Horses, mules, and cattle flee,
Donkeys break loose, struggling free.
Pigs won't eat, dogs bark with fear,
Rabbits jump, their ears stand clear.
Ducks leave water, chickens rise,
Pigeons scatter, shun the skies of home.
Sleeping snakes wake far too soon,
Rats in clusters fill the streets at noon.
Loaches twist and leeches churn,
Fish float up and slowly turn.
Dragonflies in swarms take flight,
Bees flee hives and sting in fright.

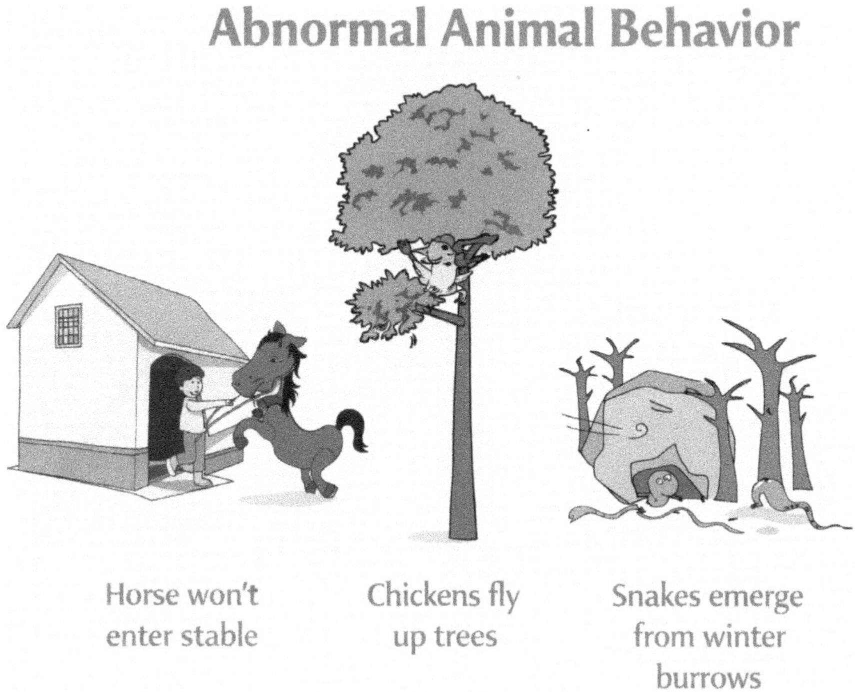

7.6 Earthquake light and sound phenomenon

Earthquake lights are glowing or flashing lights that people may see in the sky during or shortly before a major earthquake. These lights are usually blue or white, but they can also appear in other colours.

In general, small earthquakes rarely produce earthquake lights. When such lights do appear, they often suggest that a strong earthquake may occur soon. Before the 1975 Haicheng earthquake in Liaoning and the 1976 Tangshan earthquake in Hebei, earthquake lights were especially noticeable.

If people can take quick and decisive earthquake-safety actions when these warning signs appear, it may be possible to reduce harm and avoid some earthquake damage.

Chapter Eight

Earthquake Safety and Self-Protection

When an earthquake occurs, the most effective response is to stay calm and take shelter where you are. Take cover immediately during the shaking, and evacuate quickly to a safe area after the shaking stops.

Follow the steps to act immediately during an earthquake
Drop: squat or kneel on the ground to avoid falling, and keep your body as compact as possible.
Cover: move to a suitable safe spot and protect your head and neck to prevent injury to vital parts of the body.
Hold On: hold firmly onto a protective or stable object until the shaking stops.

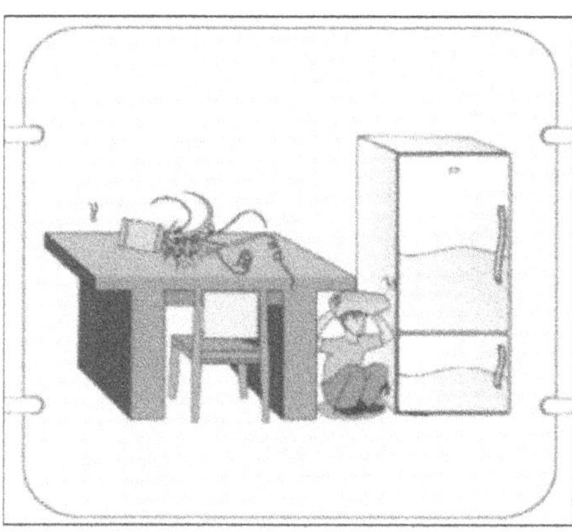

8.1 The safest body position during an earthquake

a. Squat down or sit down, curl your body as much as possible, and lower your center of gravity.
b. Hold onto sturdy objects such as table legs.
c. Use your hands, clothing, or other available items to protect your head and neck, shield your eyes, and cover your mouth and nose.

8.2 Safety at home during an earthquake

Safety Tips:

1. Hide first, then evacuate. Do not run immediately.
2. Take shelter in strong and stable places, such as under a bed, in small rooms, or in areas with structural support.
3. If time allows, open doors first and turn off gas and electricity supplies.

During an earthquake, you should take shelter nearby based on where you are at the moment, and evacuate to a safe place only after the shaking has stopped. Take cover under or beside sturdy objects that are not likely to tip over and can protect your body, such as tables or beds. You may also quickly move to rooms with smaller spaces and stronger support, such as kitchens or bathrooms. If time allows, open doors to prevent escape routes from being blocked, and turn off gas valves and electrical switches.

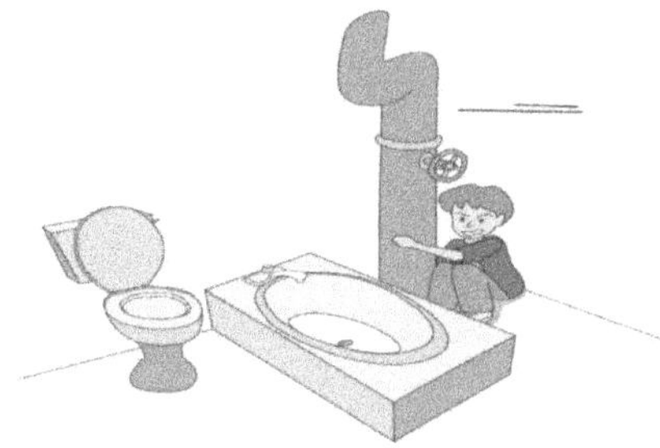

8.3 Safety in multi-storey buildings

During an earthquake, stay in safer indoor places such as under a sturdy table or beside a bed, next to low and solid furniture, or in small rooms with structural support, including bathrooms, kitchens, and storage rooms.

Quickly turn off electricity and open flames if possible. Never stay on a bed, never jump out of a building, do not go onto balconies, and do not stay near exterior walls or beside beds.

When leaving high-rise buildings, use designated safety stairways and do not use elevators. If an earthquake happens while you are in an elevator, leave as soon as possible. If the doors cannot be opened, squat down and hold the handrails firmly.

Move away from exterior walls! | Do not jump from windows! | Evacuate through safe exits.

8.4 Safety in public places

Safety Tips:

1. Follow the instructions of police officers or on-site staff and do not panic.
2. Do not rush toward exits. Keep a safe distance from the people in front of you.
3. Avoid moving with dense crowds to prevent trampling or being pushed against walls or fences.
4. If you become caught in a crowd, unbutton your collar, cross your arms in front of your chest, and protect your chest.

At schools

When an earthquake happens during class, follow the teacher's instructions. Quickly lower your head and hide under your desk to protect yourself. If evacuation is organised, leave the classroom in an orderly manner and move to a safe place. When you are on the playground or outdoors, squat down where you are and protect your head with both hands. Stay away from tall buildings and dangerous objects.

How to protect yourself during an earthquake at school

Hide under your desk | Evacuate in an orderly manner | Stay away from buildings, squat down, and cover your heac

In places such as shops, bookstores, and exhibition halls

Stay where you are and squat down near sturdy counters, pillars, or inner corners of walls. Use your hands or other objects to protect your head. Stay away from glass doors, windows, and glass display cases. Avoid shelves that are tall, unstable, or holding heavy or fragile items. Keep clear of hanging objects such as advertising boards and chandeliers.

In theatres, cinemas, and sports venues
Squat down where you are or lie under the rows of seats. Avoid hanging objects such as chandeliers and ceiling fans. Protect your head with your hands or other items. After the shaking stops, evacuate in an orderly manner under the guidance of staff.

In a moving vehicle
Hold on to handrails to avoid falling or being injured. Lower your center of gravity and stay close to your seat. Get out of the vehicle only after the earthquake has passed. Drivers should pull over to a safe area as quickly as possible and apply the handbrake to stop the vehicle safely.

8.5 Safety in outdoor areas

Avoid dangerous areas near mountains. Stay away from steep slopes, cliffs, and sharp ridges, as these areas are prone to rockfalls, landslides, and ground collapse during an earthquake.

If a collapse or rockfall occur, run to either side of the direction in which the rocks are falling. If you are already outside the affected area, go around it instead of passing through. If there is not enough time to escape, take shelter under nearby obstacles or squat in ditches or low ground, and protect your head carefully.

When a landslide happens, move away toward areas beyond both sides of the landslide, and never run in the direction of the sliding soil or rocks. Do not run downhill along the slope. Stay away from riverbanks and valleys, and look for higher, stable slopes or wide, flat ground to stay safe. Quickly judge the direction of the landslide and check whether your position is at risk. If a fast-moving landslide makes escape impossible, stay calm, watch for ground collapse, and protect yourself from falling rocks, trees, buildings, and utility poles. If the landslide moves as a whole mass, staying in place or holding onto a large tree may reduce the risk of injury.

If you are in an area at risk of debris flows, do not run up or down the gully. Instead, run toward the slopes on both sides to leave the channel or valley area. Move toward the hillsides on either side of the debris flow path, and never follow the debris flow channel upstream or downstream. Do not stay on concave slopes with steep angles and thick soil, and quickly find stable high ground. Avoid concave riverbanks at bends and narrow, low-lying convex banks. Do not take shelter beneath steep mountain slopes, as debris flows or collapses may occur.

Chapter Nine

Post-Earthquake Survival and Disaster Mitigation

9.1 Priority actions for urban residents to prevent secondary disasters

Turn off electricity and gas supplies to prevent fires. Do not light matches, candles, or turn on electrical devices, as flammable or explosive gases may be present in the air.
Do not rush to make phone calls after an earthquake, as network congestion and outage may occur and interfere with emergency response and rescue operations.

After the shaking stops, move to a safe place as soon as possible to avoid secondary injuries. Stay away from damaged buildings, narrow alleys, steep slopes, riverbanks, high-voltage power lines, and transformers. Avoid areas where glass, debris, or unstable structures may fall. If you are outdoors, remain in open areas and keep away from crowds to prevent pushing or trampling. Pay attention to aftershocks and be ready to move again if necessary. Do not return to damaged buildings, as aftershocks may cause walls, ceilings, or objects to collapse and lead to further danger. Be alert for signs of secondary disasters such as fires, gas leaks, landslides, or falling rocks. Follow instructions from adults or emergency personnel and move to designated safe areas or emergency shelters when guided to do so.

After a major earthquake, if there is an emergency shelter nearby, go to the shelter as soon as possible. Earthquake emergency shelters are usually located in parks, green spaces, school playgrounds, or public squares. These shelters are designed to provide basic support during emergency situations. They may include emergency command services, distribution of relief supplies, temporary shelters, and emergency toilets.

9.2 Self rescue

If you are trapped under debris during an earthquake, try to remain calm in the stressful situation. Take steps to help yourself stay safe. Carefully remove loose bricks, tiles, or small pieces of debris around you to create more space to move. However, do not force anything that is too heavy to move, as this may cause additional collapse. Use nearby bricks, wooden sticks, or other sturdy materials to support unstable structures and reduce the risk of being buried again during aftershocks. Do not operate indoor utilities such as electricity or water, and never use open flames. If you smell gas, toxic odors, or notice heavy dust in the air, cover your mouth and nose with a damp cloth or clothing if possible. Avoid shouting constantly for help, as this can quickly drain your strength. Conserve your energy and signal for rescue by making tapping or knocking sounds instead.

Requirements for trapped survivors

1. Maintain a strong will to survive and overcome fear.
2. If escape is not possible, try to free your hands and feet and cover your mouth and nose to prevent suffocation from smoke or dust.
3. Stay clear-headed. Do not shout loudly. Use stones, metal tools, or other objects to knock on surfaces to communicate with rescuers, and conserve your strength.

9.3 Mutual rescue

When digging to rescue people trapped under debris, it is important to keep supporting structures in place to prevent further collapse and injury. First expose the injured person's head and remove any debris from the mouth and nose to ensure clear breathing. If the person is not breathing, initiate CPR immediately. If a trapped person cannot crawl out on their own, do not pull them out by force, as this may cause further injury. In cases of suspected spinal injury, the person should be moved using a door panel or a rigid stretcher. If survivors are discovered but cannot be rescued immediately, mark their location clearly and wait for professional rescue teams to arrive.

Mutual rescue is the most effective way to reduce casualties before rescue teams reach the scene. After an earthquake, survivors in the affected area should help rescue those who are trapped. The basic principles of mutual rescue are acting quickly, identifying targets accurately, and using proper methods. Mutual rescue requires not only compassion but also scientific and careful action.

9.4 Psychological intervention after an earthquake

After an earthquake, survivors may experience various psychological reactions as a result of the disaster. If these reactions are particularly intense or persist for a prolonged period, they may develop into mental disorders. Studies indicate that the incidence of mental disorders following major disasters is about 10%–20%, while general psychological stress reactions are even more common. Over the past two decades, psychological counselling after disasters has gradually been incorporated into rescue operations worldwide as a formal component of medical assistance.

9.4.1 Common physical and psychological reactions after an earthquake

Following an earthquake, many survivors may have experienced the loss of loved ones or suffered physical injuries themselves. In addition, disaster areas often face shortages of medical care and supplies. Feelings of helplessness and an inability to cope with the situation may lead survivors to feel confused and distressed. As a result, a range of physical and psychological reactions may occur, including rapid heartbeat, elevated blood pressure, difficulty sleeping, loss of appetite, weakened digestion, emotional numbness, dulled sensations, headaches, back pain, and chest pain. These reactions are often accompanied by emotional responses such as fear, anxiety, irritability, restlessness, sadness, depression, low self-esteem, and memory problems. Some individuals may also develop harmful behaviours such as hostility, excessive drinking, smoking, or dependence on medication.

9.4.2 Psychological counselling and adjustment after an earthquake

Establishing counselling services can help guide survivors to express their emotions and share their feelings with family members and friends, allowing others to better understand their experiences. Survivors are also encouraged to talk with people who have gone through similar experiences and share their thoughts and feelings. Ensuring sufficient sleep and rest is important, and spending time with family and friends can help create a supportive emotional environment. As physical and emotional wounds gradually heal, individuals should develop a healthy outlook on life by cherishing life, appreciating its beauty, experiencing its joys, and strengthening inner resilience, while gradually restoring their daily routines to normal.

9.5 Post-Traumatic Stress Disorder (PTSD) and psychological intervention for children after earthquakes

Earthquakes are major disasters that can leave survivors, especially children, feeling helpless and overwhelmed. These experiences may lead to a range of physical, psychological, cognitive, and behavioural impairments, including sadness, tension, anxiety, fear, and distress. Therefore, timely psychological support and care are critically important.

Physical and emotional impairments may include persistent tension or insomnia, frequent fear and inappropriate reactions to the surrounding environment, exaggerated responses to stimulation, nervousness and restlessness, and developmental regression accompanied by physical symptoms such as stuttering, loss of bladder control, bedwetting, dizziness, and stomach pain.

Psychological impairments are often reflected in repeated re-experiencing of the traumatic event, emotional instability, irritability, and sudden withdrawal. At times, children may become unusually quiet or detached. They may describe unreal or imagined experiences, suffer from frequent nightmares, or repeatedly talk about disaster-related events.

Cognitive impairments may include strong feelings of guilt, helplessness, and self-blame. Some children may believe that the disaster occurred because of their own actions or mistakes. As a result, they may become depressed and, in some cases, show self-harming or aggressive behaviours.

Behavioural impairments are often characterised by reduced responsiveness to the external environment and emotional apathy. Children may appear expressionless, distracted, silent, or socially withdrawn. Their ability to manage daily activities may decline, and some may even struggle to eat or dress themselves. They may also experience difficulties with memory and concentration and show decreased participation in school activities.

Psychological crisis intervention for children in disaster-affected areas generally focuses on several key situations. During and after an earthquake, children often experience a significant loss of security. For a considerable period, they may become highly concerned about basic survival issues, such as whether their environment is safe and whether their health is protected. They may also worry about the safety of parents, relatives, classmates, and teachers, which can lead to behaviours such as panic, helplessness, avoidance, regression, and fear. Some children may even experience emotional shock, appearing dazed, slow to respond, emotionally numb, suffering from memory loss, or, in severe cases, losing consciousness.

A key approach to addressing emotional shock is to encourage emotional expression. Children should be allowed to cry or talk about their feelings freely. Supportive and accepting language should be used to comfort and guide them. For younger children or children of the same gender as the helper, gentle physical reassurance such as light touching or hugging may help increase their sense of safety. Parents can also assist in this process under the guidance of professionals.

Some children may not show serious physical injuries immediately after the disaster, but without timely psychological adjustment and intervention, they may develop psychological difficulties in the future. For children whose relatives were injured or died, especially those who lost multiple family members, the main challenge is the loss of belonging and love. Such children may find it difficult to face sudden tragic news and may experience emotional breakdowns characterized by extreme fear, uncontrollable crying, impulsive behaviour, or even self-harm or aggression.

In cases of emotional breakdown, the first step is to help the child calm down, then encourage emotional release and guide them to shift their attention to constructive concerns, such as helping injured family members. At the same time, governments, schools, family members, and friends should provide additional care and support to help children rebuild a sense of belonging and emotional security. Effective measures should also be taken to prevent accidents or harmful behaviours.

Children who suffer physical injuries may struggle to cope with their injuries or possible long-term disabilities. Some may react with distressing questions such as "Why did this happen to me?" and may develop resentment, hostility, or anger. Many of these children may later develop post-traumatic stress disorder, a delayed and long-lasting psychological disorder characterized by repeated re-experiencing of trauma, emotional irritability, and avoidance behaviours.

For such children, the first few days after the traumatic event represent the most critical period for psychological intervention, and they should receive priority psychological support. Counsellors should first help them understand what happened and clarify the uncertainties they may still have about the event, assisting them in making sense of why the incident occurred. Although learning the truth may initially cause distress, counsellors can provide emotional support and encouragement during this process, helping children gradually overcome their pain.

It is important to recognise that if this critical period for psychological intervention is missed, later treatment may still be possible, but it is generally far less effective than timely crisis intervention during the emergency phase.

During psychological crisis intervention, shortages of professional personnel and the difficulties of disaster relief sometimes result in non-professionals providing psychological support. In such cases, people may try to comfort children by concealing the truth or telling well-intentioned lies, but this approach is not beneficial for psychological recovery. Teachers should carefully observe the behaviour of traumatized students in daily school life to identify possible extreme psychological reactions. However, teachers should not attempt to provide psychological counselling without proper training and should instead seek assistance from qualified school counsellors.

9.6 Family emergency evacuation planning and practice

a. Four basic principles for evacuation
do not talk, so that evacuation instructions can be clearly heard;
do not run, to avoid injuries;
do not push or crowd others, to prevent harm to yourself or others;
do not return, to avoid further danger.

b. Four key steps for a family escape drill
Identify emergency exits
Determine one or two emergency exits in your home. If you live in a high-rise building, find the building's emergency stairways and exits in advance. Elevators must not be used during emergencies such as earthquakes or fires.

Decide on a safe meeting place
Choose a safe meeting place outside the home. During an earthquake, when there is smoke, or when a fire alarm sounds, everyone in the family should go to this location.

Escape through the nearest exit

Discuss and identify all possible exits in each room of the home and how to escape safely. Practice evacuating from different locations in the house through the nearest exit to the designated meeting place.

Do not return until safety is confirmed
Once you have reached the safe meeting place, stay there. Do not return to the house until you are informed that it is safe to do so.

In addition, families are encouraged to participate in the development and practice of community emergency plans, such as those organised by neighbourhoods or local council.

9.7 Emergency supplies preparation

If you live in an area that is prone to earthquakes, other natural disasters or plan to travel to such a region, it is recommended that you prepare or carry an emergency kit. To help ensure that your family can evacuate safely and quickly in the event of a disaster, prepare the emergency kit in advance. It is recommended to keep one emergency kit at home and another in your vehicle. It is also recommended to check the emergency supplies in the kit regularly and pay attention to the expiration dates of food, water, medications, and other items. Replace them as needed to ensure the kit is always ready for use.

Emergency Supplies to Prepare

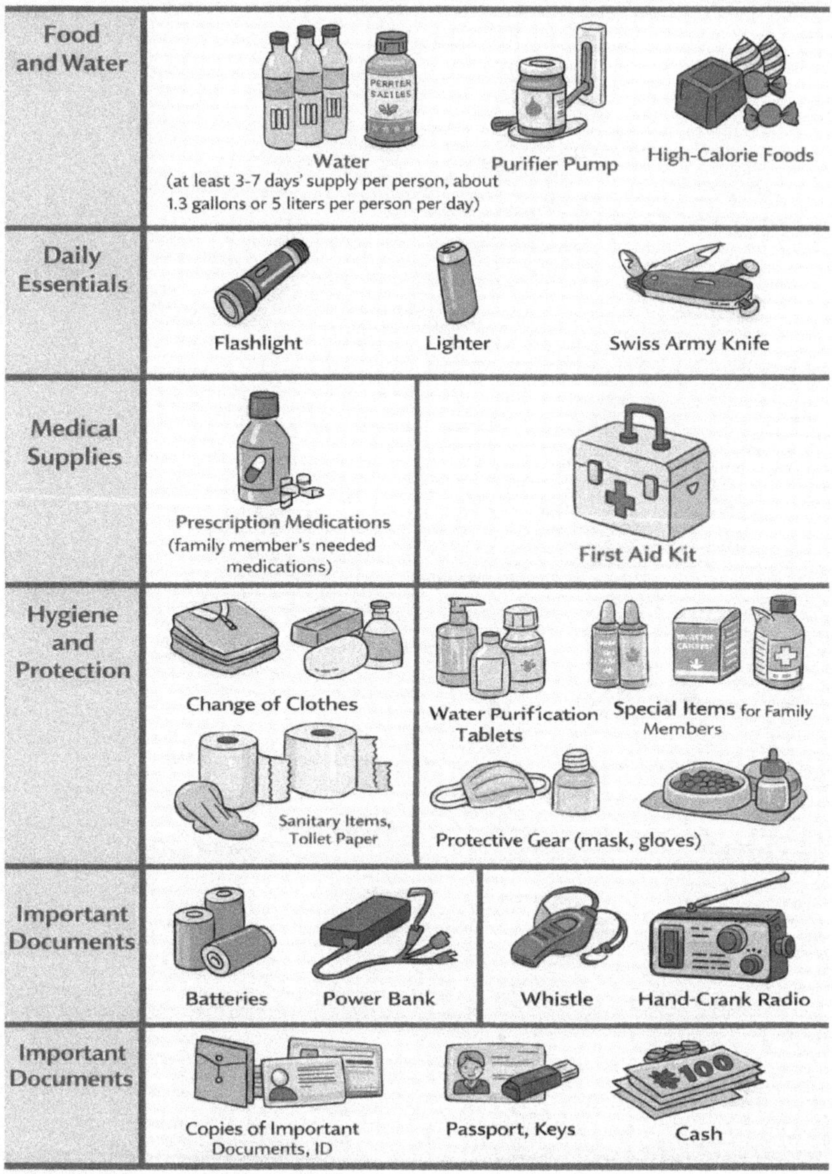

About the Chief Editor

Sun Zhuozi (born in 1947) is a survivor of the 1976 Tangshan Earthquake and a researcher dedicated to the study and promotion of earthquake culture. Having experienced the disaster firsthand, he witnessed both the devastation and the rebuilding of Tangshan. This experience profoundly shaped his long-term commitment to documenting earthquake history and promoting public awareness of disaster risk reduction and community resilience.

After retirement, Sun devoted himself to public science education and advocacy for disaster preparedness and mitigation. He has actively participated in educational activities, donated books to schools, and contributed to community outreach programs promoting earthquake awareness. Sun has authored and edited several books on Tangshan's reconstruction and earthquake culture, including The Soaring Phoenix: The Rise of Industry and Transportation in Tangshan's First Twenty Years of Reconstruction (《凤凰腾飞——新唐山二十年工交企业的崛起》, 1996) and Forty Years of New Tangshan: The Reborn People of Tangshan After the Earthquake (《新唐山四十年——大地震后重生的唐山人》, 2016). The Chinese edition of this book, The 1976 Tangshan Earthquake: Memoir and Record of Disaster Preparedness and Mitigation, has been collected by the National Library of China.

About the Editor and Translator

Casey Huang is a NAATI-certified Chinese–English interpreter and translator based in Australia. She holds a Master of Arts in Chinese Translation and Interpreting from The University of Queensland and has extensive experience providing interpreting and translation services in medical, legal, academic, and community settings. Her work focuses on promoting clear cross-cultural communication and making Chinese knowledge accessible to international audiences. For comments or feedback regarding this book, the translator can be contacted at: caseygo.translation@gmail.com

www.ingramcontent.com/pod-product-compliance
Lightning Source LLC
LaVergne TN
LVHW050935210926
848482LV00013B/620
* 9 7 8 0 6 4 6 7 3 8 7 6 5 *